Easy Home Plumbing

By Richard V. Nunn

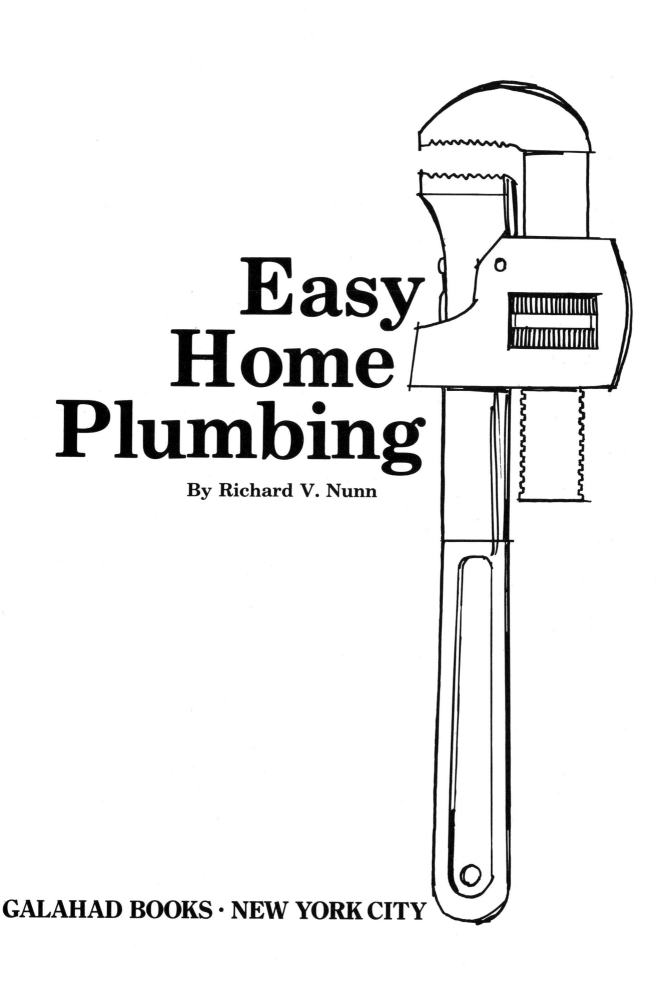

GALAHAD BOOKS · NEW YORK CITY

Published by arrangement with Oxmoor House, Inc.
ISBN: 0-88365-341-9

Copyright © 1976 by Oxmoor House, Inc.
Book Division of the Progressive Farmer Company
P.O. Box 2463, Birmingham, Alabama 35202

All rights reserved. No part of this book may be reproduced in any form or by any means without the prior written permission of the Publisher, excepting brief quotes used in connection with reviews written specifically for inclusion in a magazine or newspaper.

Library of Congress Catalog Card Number: 75-32255

Manufactured in the United States of America

First Printing 1976

Easy Home Plumbing

Editor: Candace C. Franklin
Cover Photograph: Taylor Lewis

Galahad Books believes that the information contained in this book is accurate. However, Galahad Books does not represent or guarantee that the book is free of inaccuracies, nor does Galahad Books represent or guarantee that members of the public can necessarily perform all of the operations described in the book without substantial risks to their personal safety. Readers should proceed only at their own risk in applying the techniques and advice contained in the book.

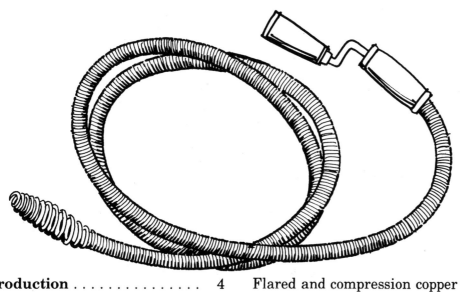

Introduction 4

Before You Start Any Plumbing Project 5
   Basic plumbing tools 5
   How the plumbing works in your home 5
   Plumbing terminology 6
   Where to turn off the water . . . 7

Common Plumbing Problems and How to Solve Them . . 9
   Quick cures for leaky faucets . . 9
   Opening clogged drains 17
   Tree roots in the sewer system 22
   Sewer gas problems 22
   Fixing leaky pipes 24
   Curing frozen pipes 25
   Stopping funny noises in pipes 32
   Disposal and hot-water heater drainage 33
   Flush tank and toilet problems 36

Galvanized Steel and Cast-Iron Pipes and Fittings 48
   Cutting and threading steel pipe 49
   Assembling steel pipe 51
   Hub-and-spigot cast-iron pipe and fittings 55
   No-hub cast-iron pipe and fittings 56
   Standard steel pipe fittings . . . 56
   Working with drainage pipe . . 57
   Troubleshooting septic systems 59

Copper Pipe and Tubing 61
   Sweat-soldered copper fittings . 63
   Flared and compression copper fittings 66

Plastic Pipe and Tubing 69
   Cutting and assembling 70

Plumbing Projects 74
   Select the proper plumbing supplies 74
   Step-by-step planning 75
   Sink and dishwasher hookup . . 83
   Sink with grease trap hookup 83
   Water supply system for one bathroom and a kitchen 84
   Two ways to join a new drainage system to an old one 84
   Reinforcing a cut joist with headers 85
   Horizontal pipes through joists 85
   Running pipes through walls and floors 86
   Framing and measurements for a lavatory 88
   Framing for a bathtub 89
   Framing for a floor-mounted toilet 90
   Vent stack detail 90
   Locating an opening for a soil stack 91
   Typical waste and vent run . . . 91
   Connecting lavatory waste pipes 92
   Two ways to connect a toilet tank 93
   Measuring lengths of nipples . . 94
   Setting a toilet bowl on its flange 94

Index 95

Introduction

There is an aura of mystery surrounding plumbing repairs that worries even the most daring do-it-yourselfers.

The purpose of *Easy Home Plumbing* is to solve this so-called "mystery" and to put you, the homeowner or apartment dweller, in a position to deal with most plumbing emergencies in your home and in doing so to help you save a bundle of money.

First, let's put your fears to rest.

The plumbing system in your home consists of an incoming water supply and outgoing drainage pipes. The incoming water supply includes one series of pipes for cold water and a second series of pipes for the hot water. The series of outgoing drainage pipes handles water removal, waste removal, and acts as a vent for unpleasant, harmful odors in the waste system.

When the plumbing system malfunctions, the problem will usually be one of these:
- Clogged pipes or drains
- Leaking pipes and/or valves

When you're adding more plumbing to your home, you will be involved in assembling the following:
- Cold water pipes
- Hot water pipes
- Waste pipes

Don't be deceived. *Easy Home Plumbing* is not a training manual for would-be professional plumbers. The book is designed to show you how to meet common plumbing breakdowns. It includes instructions on how to work with galvanized steel pipe, copper tubing, and plastic pipe, and, as a bonus, you will find a potpourri of simple plumbing projects that you can do when remodeling or adding extra space to your home.

Before You Start Any Plumbing Project...

There are two ironclad rules you must follow when attempting any plumbing repair or remodeling job:
1. Turn off the water at the main shutoff valve. Or turn off the water at the individual shutoff valves located below each sink, lavatory, or flush tank, depending upon where repairs need to be made.

    The hot and cold water pipes in your home are under pressure—usually 50 to 60 pounds per square inch. If you unscrew a coupling on a pipe or remove the stem assembly from a faucet and the water has not been turned off, you'll quickly find yourself being flooded.

    Waste pipes are not under pressure; they operate on a gravity principle. The waste flows down and out of the system into the sewer (or septic tank) lines.
2. Plumbing repairs—and plumbing remodeling—must comply with the building and plumbing codes in your town or city. The dealer from whom you purchase your plumbing supplies and accessories will probably know codes. If he does not, check the city building department, city hall, or the county courthouse for the information.

Basic plumbing tools

For basic plumbing repairs and maintenance, you will need these tools:

Auger (plumber's snake) ($3 to $5)
Plunger (plumber's friend) ($1.25)
Two 10-inch pipe wrenches (about $8.00 each)
4-inch adjustable wrench ($4 to $6)
Assortment of plumbing washers ($1.50)
Standard blade screwdriver ($2)
Phillips screwdriver ($2)
Slip-joint pliers ($1 to $2)
Needle-nose pliers ($4)

For extensive repair jobs and for adding new plumbing, you will need these tools also:

Replacement faucet seats (cost depends on size, type)
Compression nuts and rings (cost depends on size, type)
Drain cleaner ($1.50)
Faucet seat dresser ($5)
Solder and flux ($3)
Propane torch ($10 to $20, depending on accessories)
Graphite packing ($1)
Tube cutter ($3)
Electrician's tape ($1)

For cutting threads on galvanized pipe or for flaring copper tubing for compression-type fittings, we suggest that you rent the proper equipment since the cost of these tools may be prohibitive if you plan to use them only one or two times. The equipment includes a vise, threading dies, and flaring block.

How the plumbing works in your home

The water in your home is probably supplied by a public utility. If not, the water comes from a well or spring in the ground.

The incoming water-supply pipe divides at the hot-water heater into two separate, continuous systems: one for cold water and one for hot water. You can trace the hot and cold water pipes starting at the water heater. The hot water pipe is always on the left side of the heater; the cold water pipe is on the right. Generally, the pipes run parallel to each other throughout the house.

As pointed out earlier, hot and cold water are under pressure, which is how the water is able to flow upward in your home.

Compared to the pipes in the drainage system, water-supply pipes are relatively small. Such pipes measure anywhere from $3/8$ to 1 inch inside diameter. The amount of water that comes out of the faucets in your home depends on the diameter of the supply pipe, the length of the supply pipe, and the height the water has to rise in your home. For example, the water in a second story bathroom may not run as freely as the water in the kitchen or basement fixtures. Slow water flow may also be caused by too many outlets on the pipe, or slow water may result when the fixture is located a long distance from the water main. The only solution to these problems is larger pipes.

The drainage system is completely separate from the incoming water-supply system. It handles drain water removal, waste removal, and odor and gas removal (venting).

The "soil stack" is probably the largest pipe in your home. It extends vertically from the lowest part of the house or plumbing system up through the roof. The soil stack is connected to the main sewer pipe (or a septic system) and the stack usually has a cleanout plug. It collects waste from several fixtures including the toilet.

Another purpose of the soil stack is to vent odors from the fixtures. The stack is vented through the roof since all gases rise—even through the same pipes that carry waste down. Soil stacks may be copper, cast iron, or plastic pipe, depending on the plumbing codes in the area.

The waste removal pipes in the drainage system are $1\frac{1}{2}$ or 2 inches in diameter. These pipes accept the discharge fluids from all fixtures in your home (except the toilet which has its own soil pipes for waste removal). Drain water removal pipes are often hooked into the soil stack, in order to vent gasses. However, sometimes drain water removal pipes are vented separately through the roof.

"Vent" pipes equalize the pressure between the fixture's trap and the waste pipe. They permit water and waste to go through the waste pipes without clogging. The vents also carry off sewer gas and prevent waste from backing into a fixture that is located below another fixture.

Drainpipes at each fixture in your home pass through an S- or U-shaped curve called a "trap." Traps in the plumbing system provide a liquid seal by retaining water between the fixture and the waste pipe. The purpose of the seal is to keep air from entering the waste pipe while water is flowing. The seal prevents gas and other odors from entering your home from the main sewer lines.

Plumbing terminology

When dealing with a plumbing supply store, a professional plumber, or even a manufacturer's instruction booklet, there are several plumbing terms you should know which can save you time and money.

Branches are hot or cold water pipes or waste pipes that are connected to the various fixtures.

Fixtures are lavatories, toilets, sinks, bathtubs, and showers to which fittings (faucets, valves, traps) are connected.

Sewer main is the big pipe that funnels the waste from your home to the city disposal plant or to the septic system.

Risers are supply pipes that extend to the second or third floor of your home (higher if you live in a high-rise apartment building).

Valves are faucets or shutoff valves.

Where to turn off the water

The main water valve shuts off all the water in your home. This valve is usually located near the water meter or near the hot-water heater. *Before making any plumbing repairs you must turn off the water at the main shutoff valve or at the particular lavatory, sink, or toilet shutoff valves.*

It is advisable to test the main shutoff valve before any trouble arises. Turn off the water; then turn on a faucet in the house. If water continues to run out of the faucet, you have not turned off the main valve. Once you have located the main valve, label it clearly so you will know where to go to turn off the water in the event of an emergency—or to carry out a repair.

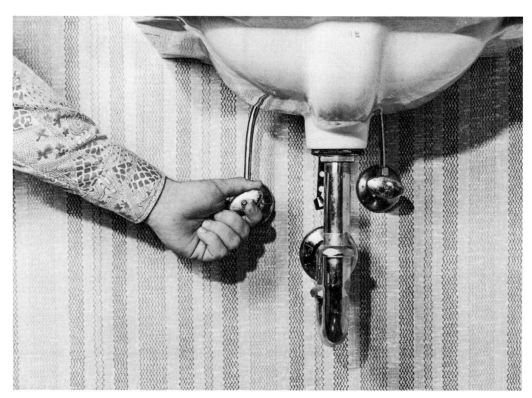

Shutoff valves for individual fixtures such as lavatories, sinks, and toilet flush tanks are located directly under the fixture where the water supply is connected. If your home's plumbing is equipped with these shutoff valves, you do not have to turn off the water at the main valve to make repairs. But before you disassemble a faucet or a flush tank valve, turn off the water at the shutoff valve, and open the faucet or flush the toilet to make sure the shutoff valve is working. If it is not, turn off the water at the main valve.

Before You Start

The plumbing system in your home looks something like this. The hot and cold water pipes are continuous and join at the hot water heater. Since the hot and cold waters are under pressure, the pipes may be horizontal or vertical. Drainage pipes, which are not under pressure, have a slight downward pitch to take advantage of the pull of gravity.

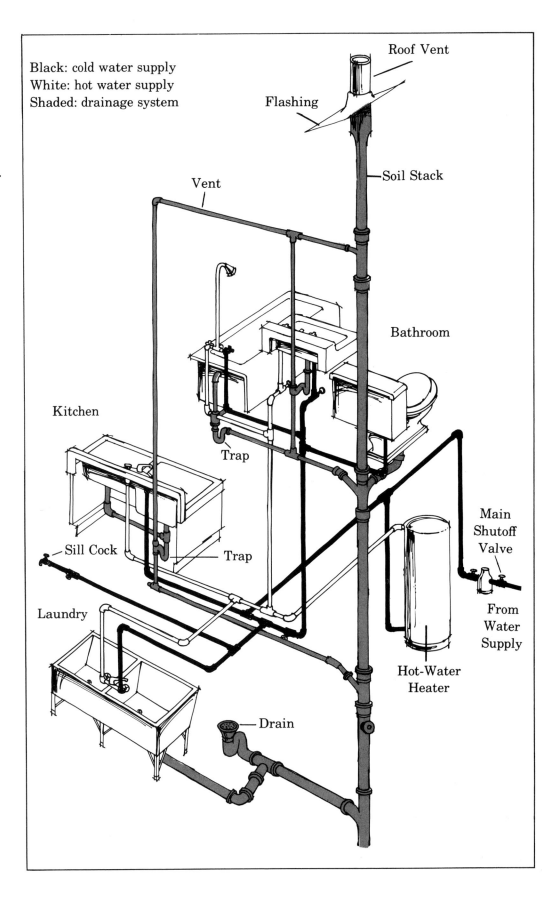

Black: cold water supply
White: hot water supply
Shaded: drainage system

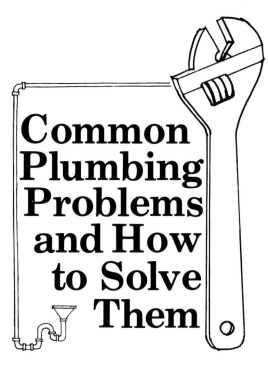

Common Plumbing Problems and How to Solve Them

Sooner or later the faucets in your home will develop the sniffles and drip. It is also likely that a flush tank will go on the fritz—either an obstruction will block the toilet, the water will run constantly, or you will hear splashing sounds inside the tank.

These conditions are quite common, so don't panic when problems arise. Pipes can become clogged with soap, hair, grease, and other objects such as hairpins, diapers, toys, and perhaps lime deposits from the water itself. Valves and washers wear out simply because of old age. Washers are rubber or a plastic material that will normally wear out after a period of time. Pipes can leak because of poor connections or simple corrosion.

Plumbing maintenance can save you plenty of grief in the case of clogged pipes and lime deposits. Give drains a weekly dose of commercial drain cleaner to cut away deposits of grease and hair in the pipes. This includes lavatory and sink drains as well as bathtub drains. If you have a garbage disposal in your home, make sure you run plenty of water through the unit while it is in operation. The water flushes away grease, which can cause blockage in the pipes below.

If lime is a problem in your area, you can sometimes prolong the life of the plumbing pipes by installing a water softener.

If the pipes in your home are already clogged with lime deposits and water is flowing slowly through the pipes, the pipes will probably have to be replaced. Do not attempt to pour a chemical through the pipes to clean them; the resulting chemical reactions can be very dangerous. If the water flows slowly and lime deposits are not evident in the pipes (you can tell by unscrewing a pipe connection after you turn off the water), the problem may be restricted flow at the meter. Call the water utility serviceman to check the water supply at the meter.

Quick cures for leaky faucets

In general, faucets leak because of worn washers or a worn faucet seat, which lets the water trickle past the washer seal. Or a faucet may leak because the cap that holds the stem assembly in its housing is not screwed down tightly or because the graphite packing or washer in the stem is worn.

If you have a chronically leaking faucet in your home, you might consider replacing it rather than continually trying to repair it. You will find that replacing the faucet is almost as easy as replacing a washer, although the new faucet will cost you more than the simple washer replacement.

For less than $2 you can buy a package of assorted flat washers and O-ring washers. Often the bag also contains little brass screws which are sometimes needed to replace corroded screws that hold the washers on the bottom of the stem assembly.

You can purchase assorted sizes and types of washers at plumbing supply shops, building material retailers, hardware stores, and most large department stores.

To change a washer you will need a screwdriver (standard blade or Phillips), an adjustable wrench with smooth jaws (or channel lock pliers), and a wad of fine steel wool.

Remove the faucet handle. The handle may have a decorative insert that hides the screw. If so, pry off the insert with the blade of a screwdriver or knife. The screw that holds the handle to the stem assembly may be a standard slotted screw or a Phillips screw. If the screw is a standard one, make sure the blade of the screwdriver fits the slot. If it does not and if you use force on the blade to turn the screw, you will not only mar the finish on the handle, but you will also "chew up" the screw head.

Pry off the faucet handle. Pad the screwdriver with a cloth or adhesive bandage, to prevent the screwdriver from marring the faucet's finish. If the handle doesn't want to come off, try working the screwdriver around the base to loosen the handle on the stem assembly. Go easily; you don't want to bend the handle housing.

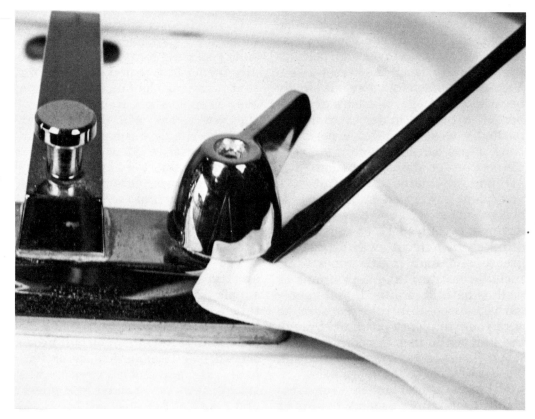

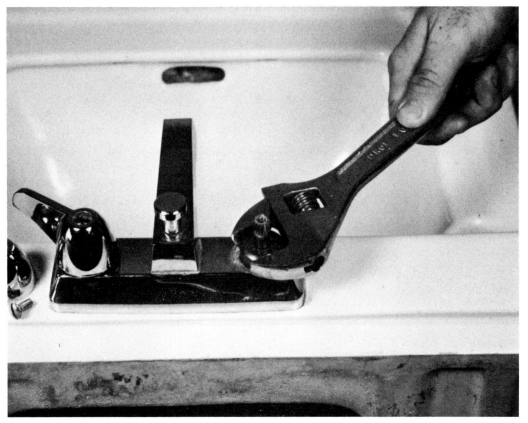

Loosen the screw cap that holds the stem assembly in place. A smooth-jawed adjustable wrench is best for this. If you don't have an adjustable wrench, use channel-lock pliers, regular pliers, or vise-grip pliers. But pad the jaws of these pliers so that you don't mar the cap fitting. Once the screw cap has been loosened, turn out the stem assembly with your fingers.

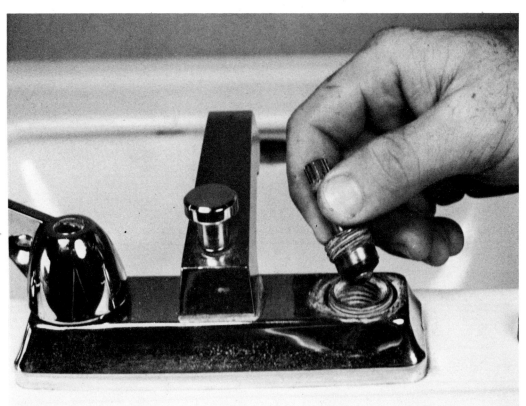

Twist out the stem assembly. If the stem is difficult to remove, replace the handle on the stem (without the screw) and twist it out. The stem is threaded, as shown, and will make several turns before it is free of the housing. Do not risk damaging the stem by forcing it out of the housing with pliers.

With fine steel wool lightly buff the stem assembly to remove any corrosion. When you are finished, the stem should be shiny clean. At the bottom of the stem is the washer which is held in position by a tiny, usually brass, screw.

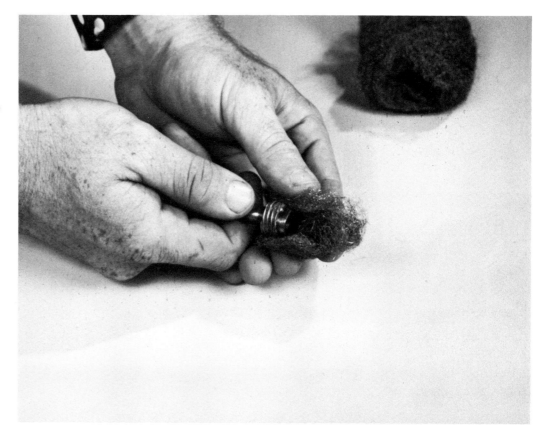

Remove the screw that holds the washer to the stem assembly. A washer needs to be replaced if it is deeply impressed and flattened and/or the edges are ragged, showing signs of wear. Throw the old washer away so you won't mistakenly use it again. Pick a washer that fits the stem out of the assortment. The fit should be a tight one. Reinsert the small brass screw. If the screw is worn or corroded, replace it as well. In an emergency, if you don't have new washers, you may be able to stop the leak by removing the old washer from the stem assembly, turning the washer over, and reseating the washer with the screw. However, you should replace the washer with a new one as soon as possible.

A damaged valve seat as well as a damaged washer can cause a faucet to leak. You can usually tell if the seat is "rough" or pitted by the condition of the washer on the end of the stem assembly. If the washer is really chewed and ragged, chances are the valve seat needs "dressing" or smoothing. A special valve seat grinding tool, shown here, is used for this. It has a grinding die that fits into the housing of the faucet and smooths any roughness in the valve.

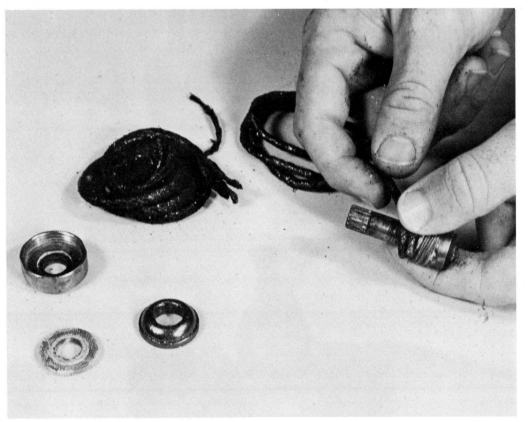

Damaged packing around the cap of the stem assembly can cause a faucet to leak. First try tightening the cap as shown. Don't use a lot of pressure or you may damage the cap and stem assembly. If tightening does not stop the leak, remove the cap. Then, with a nail, ice pick, or brad awl, remove the impacted material inside the cap. Packing is available at plumbing supply outlets and most hardware stores. Wind the new packing around the stem, filling the hollows of the cap. Reassemble the faucet. Do not use regular string for packing; if you do, you will have to replace the string before too long.

Single-handle faucets are washerless. However, these faucets may have worn O-rings that are causing the leaking. To disassemble a single-handle faucet, pry off the escutcheon plate with a screwdriver or old case knife as shown. There are several parts to single-handle faucets. As you remove parts from the faucet housing, be sure to note the sequence in which the parts are assembled in order to reassemble the parts correctly.

The cap screw releases the handle which fits into a slot in the stem assembly. After the handle is removed, pull out the cartridge. This will expose the O-ring washers. Slip new O-rings over the assembly, then reassemble the faucet. If the faucet still leaks, disassemble it again, take the cartridge to a plumbing supply outlet, and buy a replacement kit to match the assembly.

The faucet assembly pulls or twists out of the housing after you remove a retainer ring, as shown. Pull the ring straight out of the stem assembly so that you don't bend it. If you do not have needle-nose pliers for this job, use a finishing nail or an ice pick. If the assembly is difficult to pull out of its housing, replace the handle on the end of the stem for better leverage.

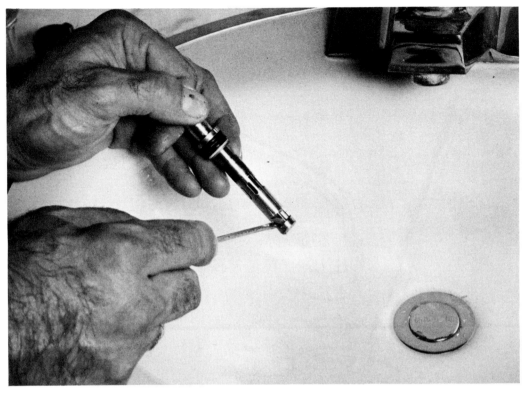

The strainer at the tip of some washerless faucets may be clogged, causing leaks or slow running water. If you find lime or lint in the strainer, clean out the debris, reassemble the faucet, and turn on the water. If the faucet is still sluggish or leaky, replace the entire cartridge assembly.

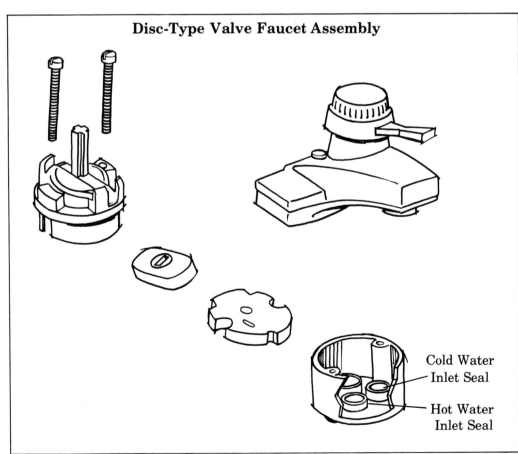

Disc-Type Valve Faucet Assembly

Cold Water Inlet Seal
Hot Water Inlet Seal

Disc-type valve faucets have hot and cold water inlet seals which may cause problems. Disassemble the faucet by removing the two screws that hold the unit together; lift out the parts, and replace the rubber inlet seals located at the bottom of the assembly.

Common Plumbing Problems 15

Slow-running faucets (usually in the kitchen) may be caused by clogged strainers. Remove the spout and escutcheon to reach the plugs and strainers which are located on either side of the faucet body. Remove the plugs and clean away any sediment from the strainers. Replace any worn O-rings.

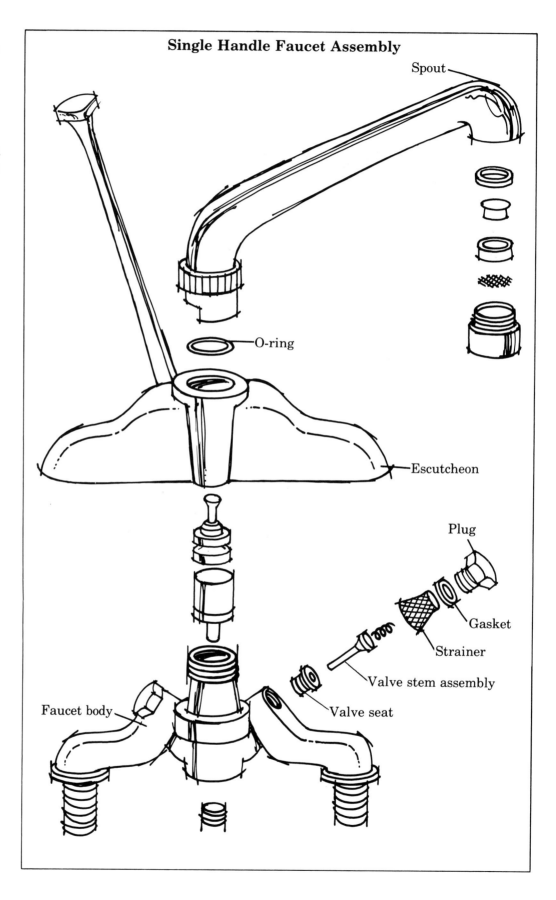

16 *Common Plumbing Problems*

Opening clogged drains

Grease and hair are the primary causes of clogged drains both in the kitchen and in bathroom lavatories and tubs.

Chemical drain cleaners help keep drains clear, and such cleaners should be applied to drains weekly. However, do not put chemical drain cleaners in the garbage disposal or in pipes which contain built-up lime deposits.

Avoid excessive grease from clogging kitchen drains by wiping away excess fat, oil, and grease from dishes, pots, and pans before washing them.

If you must wash grease down the sink, plug the opening with a stopper, then fill the sink full of *hot* water. When the sink is full of hot water, remove the stopper and let the water run out as quickly as possible. Repeat the process until the water flows out of the sink at a normal rate.

If a sink is completely clogged with debris and there is no flow at all, a chemical drain cleaner probably will not open the sink. Use a plunger or an auger to open the drain. See photographs for instructions.

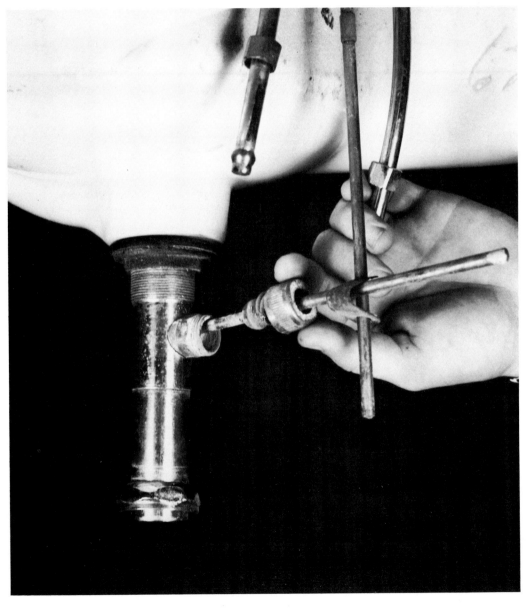

Lavatory stoppers can cause slow drainage when the stoppers become clogged with hair and soap. Pop-up stoppers are controlled by a lever which connects to the lavatory drain as shown. To remove the stopper, disconnect the lever mechanism by simply unscrewing the nut that holds the lever in position and pulling out the lever; then pull the stopper out of the lavatory.

Clean the stopper with steel wool, wiping away all hair, grease, and corrosion. Then replace the stopper in the lavatory and reconnect it to the lever assembly. For strainer type discs—usually found in kitchen sinks and bathtubs—pry out the strainer with a screwdriver or an old case knife. Remove any debris (hair and grease) with steel wool, or soak the strainer in detergent and hot water until the debris lifts off easily.

Twist-out stoppers in some lavatories are not attached to any lever mechanism. To remove the stopper simply twist it out of the drain with your fingers. Clean away all debris and replace stopper in the drain by twisting it into position.

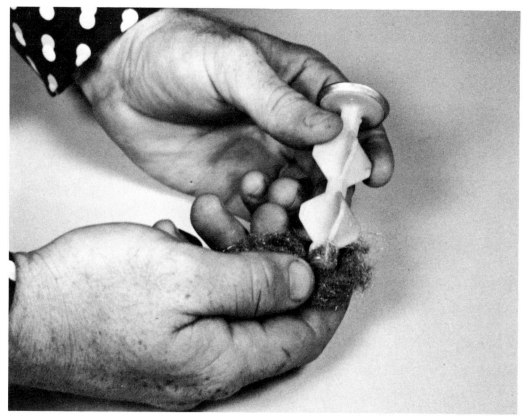

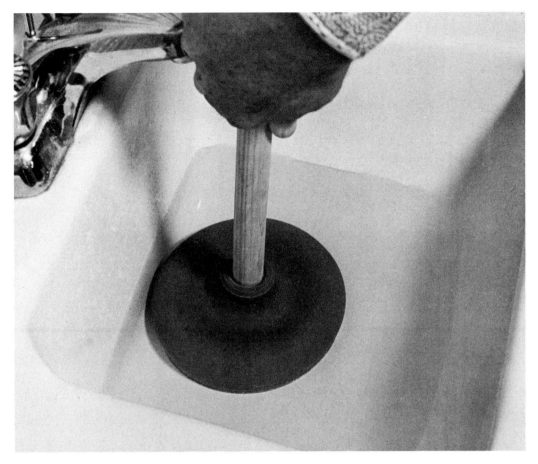

A plunger, or plumber's friend, is a large suction cup on the end of a handle. Use a plunger when the sink is completely blocked and the water does not drain out at all. First, remove the plug or stopper from the sink. Fill the sink so that it is about half-full of water. If the sink is not fully blocked (there is some water flow), turn on the cold water faucet to keep the water level in the fixture half-full. This helps seal the edges of the plunger and gives the plunger vacuum more vacuum effect.

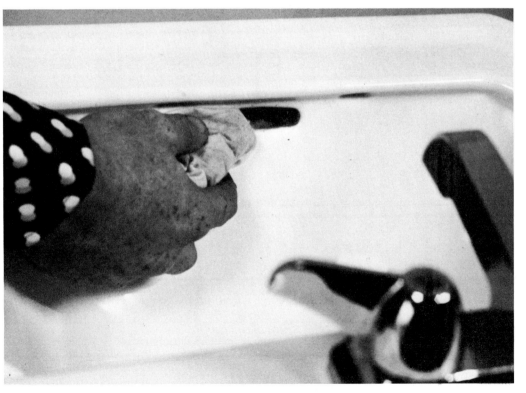

Plug the overflow outlet with a wet cloth before using a plunger. This gives the plunger more suction. If the plunger does not seem to work at first, don't give up. Keep working the plunger up and down over the drain. The plunger has to lift the clogged material within the pipe. If the plunger is not successful after repeated tries, check other drains in the house. Sometimes what appears to be a clogged drain in one fixture can actually be caused by a problem in fixtures or pipes just below that particular sink. Start looking for the trouble at the fixture that is the lowest on the drainage system.

If the plunger does not work, try an auger, or plumber's snake, in the drain. Thread the flexible rod or tape down through the trap of the drain (shown here disassembled so that you can see how the rod works). As you push the auger through the drain, turn it 360 degrees. Continue the process, even when water in the sink starts to flow out. Turn on both hot and cold water faucets for more flushing power. When the drain is open, remove the auger and give the drain a dose of chemical cleaner.

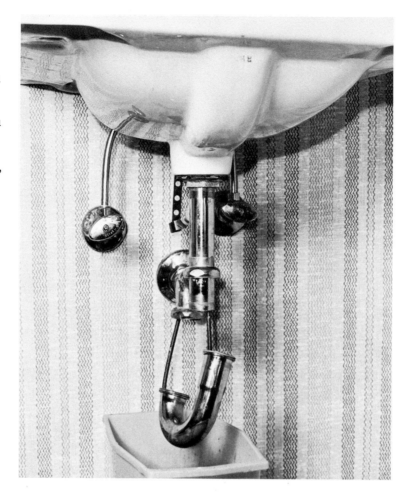

If the auger does not work, you must remove the trap. Place a bucket under the trap to catch the water in the sink, then use an adjustable wrench to turn the large nuts on the trap. If you don't have an adjustable wrench big enough to fit the trap nuts, use a pipe wrench, but pad the jaws of the wrench with masking tape or make-up pads. This will prevent the ridges in the wrench's jaws from damaging the pipe's finish.

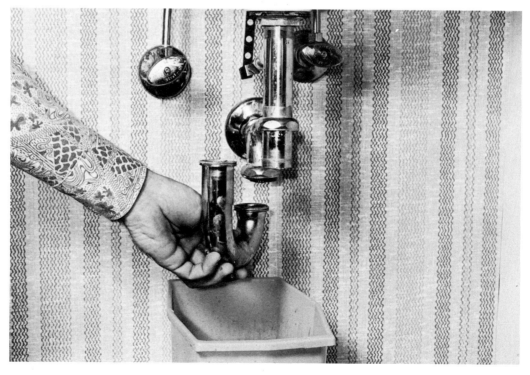

Bathtub traps are similar to sink and lavatory traps. If the trap is blocked, first try opening it with a plunger. Block the overflow outlet with a wet cloth for more suction. Doesn't work? Try running an auger into the trap opening. This is sometimes difficult to do since the pipe bends in the trap are tight. If this treatment doesn't work, remove the plate from the overflow outlet by unscrewing it as shown.

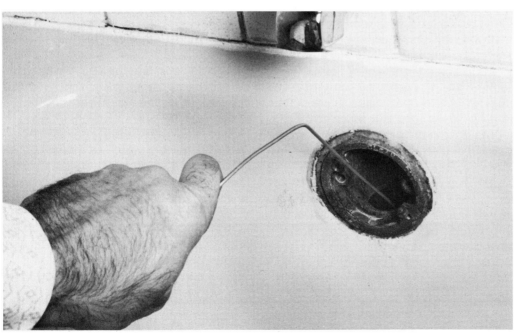

Straighten a coat hanger and insert it down the overflow outlet. A sharp hook formed on the end of the hanger with pliers will help snag any debris such as hair. For shower and basement drains, unscrew the strainer plate, pry it out, and clean the drain with an auger or length of coat hanger. When the drain is open, give it a dose of chemical cleaner.

Common Plumbing Problems 21

If neither an auger nor a plunger can open the drain, the cleanout plug where waste leaves the house may be blocked. With a pipe wrench, carefully open the cleanout plug. Have a large bucket underneath the drain to catch any waste. Then insert an auger or garden hose in the drain. Thread the auger into the pipe a short distance; then turn the crank handle and work the auger through the pipe. If you're using a garden hose, turn on the water and inch the hose through the pipe as far as possible.

Tree roots in the sewer system

If you own an older home, the sewer pipes are probably made of clay. Over a period of years these pipes may settle into the ground, causing the pipe joints to open slightly and tree roots to grow into the pipes.

You can rent a special root-cutting auger, powered by a motor, to slice away these roots. The head of the auger is equipped with special razor-sharp cutting blades. Thread the auger into the cleanout plug, turn it on, and slowly force the auger down into the pipe.

You know you have reached the main sewer line when the auger can't be moved any farther. Slowly remove the auger from the pipe. Insert a garden hose into the cleanout plug, and thread it down the pipe. Turn on the water full blast. The water from the hose helps "flood" debris out of the pipe and into the main sewer line. Close the cleanout plug and flush all the toilets in the house several times.

Sewer gas problems

If you have a toilet or drain that is seldom used, it may lose its trap water through evaporation. This allows sewer gas to escape into the house.

You can prevent this problem by filling the trap. Kerosene or oil floated on the water in the trap will slow the water's evaporation.

If this doesn't work, chances are the system is improperly vented and needs to be redesigned—a job for a professional plumber.

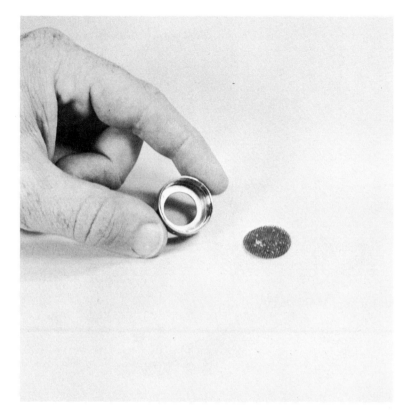

A clogged faucet aerator can result in low water pressure. Aerators are located on spigots of faucets in the kitchen and bathrooms. Remove the aerator with pliers padded so that the ridges in the jaws of the pliers don't scratch the finish on the aerator. Sometimes you can remove the aerator by simply turning it with your fingers. Remove any sediment from the screen in the aerator; then replace the unit on the faucet. If the water is still running slowly, have your local utility serviceman check the water meter and the pipe that leads directly out of it into your home. This service is usually free.

Sediment in the screen or in the tiny holes of a shower head can cause clogging and low water pressure. If the shower head is metal, rust or corrosion may be the problem. If you are unable to clean out the debris, you will have to replace the nozzle. If the water comes out of the nozzle in a stream instead of a spray, you may be able to adjust the nozzle by twisting the inside ring of the nozzle clockwise or counterclockwise.

Common Plumbing Problems

Pipes become clogged with lime because of hard water. When the lime inside the pipe becomes thick enough, the water pressure drops or stops completely. The best cure is to have a water softener installed in your home before the pipes become clogged. Sediment may be flushed from the pipes, if the liming is not too bad, but this is a job for a professional plumber. Do not try to flush the pipes yourself.

Fixing leaky pipes

Pipes generally spring leaks at connections. This may be due to rust or corrosion on the connection, the house's settling and breaking the connection seal, or freezing and thawing of the plumbing lines.

If the pipe is galvanized steel and leaking at a joint, try tightening the joint with one pipe wrench on the joint and another pipe wrench on the pipe. Don't use too much pressure; just try to turn the threads on the pipe 1/4 inch or so. Always double-wrench this job keeping the jaws of the wrenches square on the pipe and fitting. Jaws on pipe wrenches are designed to take up slack which prevents crushing the pipe.

If the joint still leaks, disassemble the pipe from the connection if the pipe has a union joint in it. (See the chapter on working with galvanized pipe.)

Remove the leaking pipe from the connection, clean the threads with a wire brush or steel wool, daub pipe compound on the threads, and reassemble the pipe in the connection.

If the pipe does not have a union fitting, you will have to cut a section from the pipe, thread the cut end, and install a new section with a union coupling. Instructions for this procedure are shown in the chapter on working with galvanized pipe.

If the pipe is copper tubing and the tubing is leaking at a joint, turn off the water at the main valve shutoff or the valve nearest the leak. Drain the tubing by breaking the soldered joint. Clean the tubing ends with steel wool and insert a new connection on the tubing. Then "sweat" the joint with solder as explained in the chapter on working with copper tubing.

If the leak is in the run of copper tubing, turn off the water at the main shutoff valve, and cut out the damaged section of tubing with a hacksaw or tube cutter. Replace the damaged tubing with a short piece of tubing and two couplings.

If the pipe is plastic, simply cut out

the damaged part of the pipe and reassemble the joints and pipe. (See the chapter on working with plastic pipe.)

You can buy patches for leaking plumbing. Such patches are for emergency use only; you should replace or repair any pipe or connection that is leaking just as soon as possible. Several types of emergency patches are pictured.

Curing frozen pipes

Pipes generally freeze because they are not properly insulated. Such pipes are usually located at sill cocks (outside faucets where you attach a garden hose), in unheated and uninsulated crawl spaces, along outside walls that do not have insulation in them, and at floor level where the pipe turns upward toward a bathroom.

There are several ways to thaw frozen pipes. The real solution, however, is to insulate pipe runs to keep the pipes from freezing.

• In open pipe runs where pipes tend to freeze, insulate the pipes with an insulating material. Insulation is available in strip form or in molded form. Molded insulation includes two halves that fit around the pipe with wire or metal strips to hold the halves together.

• If the pipe is behind the wall, and the wall is not insulated, a quick remedy is to cut a hole in the top of the wall from the inside and pour mica-type insulation into the wall. The small mica pieces will sift around the pipe. If the pipe freezes again after this treatment, you will have to open the wall and properly insulate it.

• By leaving the door open between a heated and unheated area on a very cold day you may prevent the pipes in the unheated area from freezing.

• A 100-watt light bulb placed next to an uninsulated pipe will sometimes prevent the pipe from freezing.

• Sill cocks are usually installed with inside shutoff valves which should be turned off during cold weather. If, for any reason, you want to use a sill cock during cold weather, be sure it is designed with a special long shank unit. This assembly places the sill cock outside but shuts off the water inside where it is warm.

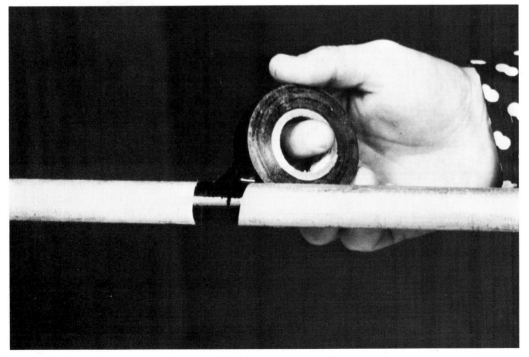

Plastic electrician's tape sometimes stops small leaks in pipe runs. Simply wrap the tape tightly around the pipe, overlapping the tape about two-thirds its width. At pipe joints, you can stop leaks by using an epoxy cement which is available at plumbing supply outlets.

A pipe clamp has a rubber gasket liner (far right) to stop leaks. Fit the gasket over the leak, place the clamp over the gasket, and tighten the bolts that hold the clamp together. Pipe clamps are available in several sizes; be sure to check the size of the pipe you are working with. Clamps are available at plumbing supply outlets and some general merchandise stores.

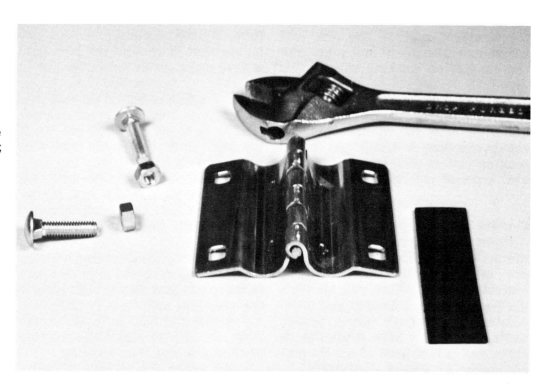

Use an adjustable wrench to tighten the bolts that hold the pipe clamp together in a clam shell configuration. The clamp should be centered over the gasket, if possible. If the pipe has a series of pinhole leaks in it, you should not attempt to clamp the leaks. Corrosion has weakened the walls of the pipe, and you should replace the entire pipe.

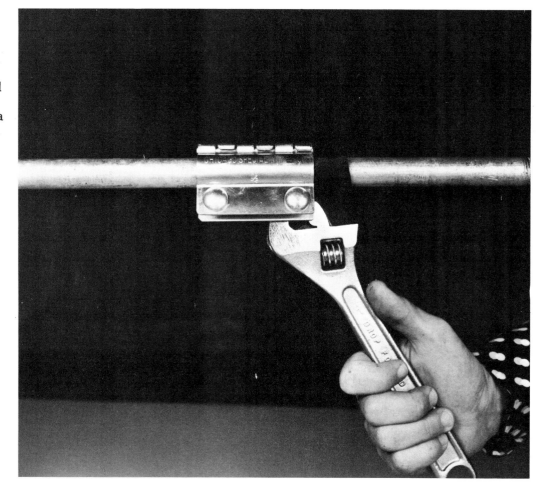

A piece of rubber or an auto hose clamp may temporarily prevent a pipe from leaking. Wrap the rubber around the pipe; then open the clamp and place it around the rubber. Tighten the screw which takes up slack in the clamp.

A C-clamp and rubber gasket will work in an emergency. Place the gasket over the break in the pipe and tighten the clamp over the gasket as shown. Remember: before you apply any clamping device or epoxy cement to a leaking pipe, turn off the water at the main shutoff valve. This reduces the pressure on the pipe and the leak.

Common Plumbing Problems

Big holes in pipes sometimes can be plugged with a self-tapping screw and soft rubber gasket. First turn off the water. Drive the screw, with gasket in place, into the pipe. You will need an adjustable wrench to drive the square-headed screw. This emergency measure is confined to fairly large pipes since the shank of the screw can block small pipes. If a large pipe—such as a drainpipe at the low end of the water system— breaks, turn off the water and call in a professional plumber immediately.

Frozen pipes are most easily thawed with electrical heating tape. Simply wrap the tape around the pipe as shown. One end of the tape has an electrical plug; insert it into a standard electrical outlet.

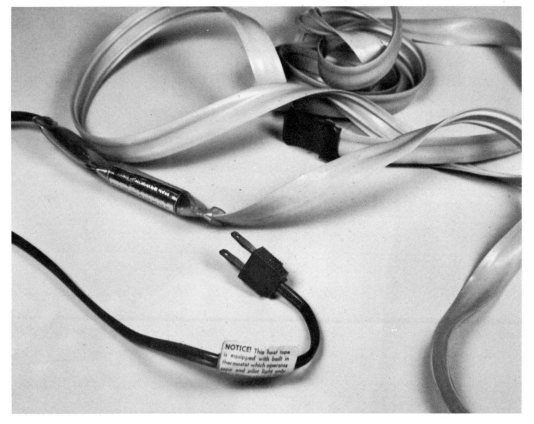

Wrap pipes which are constantly subjected to cold and freezing with electrical tape that has a thermostat on it. When the temperature drops to a certain point, the tape heats the pipe and prevents it from freezing. If you use the tape to thaw the pipe, and the tape does not have a thermostat on it, remove the tape just as soon as the pipe has thawed.

A propane torch may be used to thaw frozen pipes. Work from an open faucet toward the frozen section. Otherwise, steam created within the pipe from the torch may cause damage to the plumbing. If the frozen pipe is behind a wall, use a heat lamp to thaw the pipe. Keep the lamp away from the wall so that you don't scorch the paint or wall covering.

Common Plumbing Problems

Wrap open pipe runs with insulation to prevent the pipes from freezing and dripping with water condensation. This system involves a narrow width of fiber glass insulation and a reflective binding tape. Other insulation products available include plastic foam, wool felt, asbestos tape, pipe jackets formed from foam or fiber glass, and air-cell asbestos.

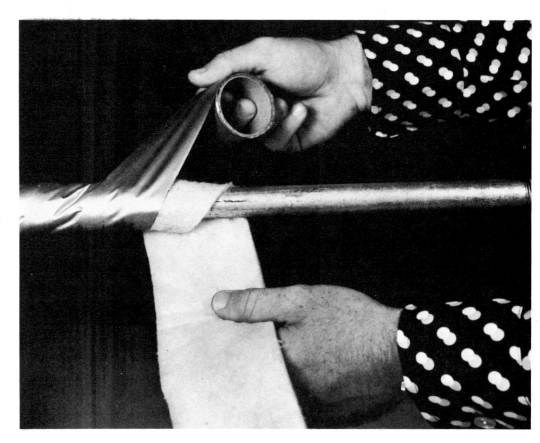

No-drip tape is used to wrap pipes plagued with condensation dripping problems. The material is dough-like in consistency. To install the tape, first wrap the tape loosely around the pipe in a spiral-like pattern. Work with relatively short lengths of the material to insure a smooth application.

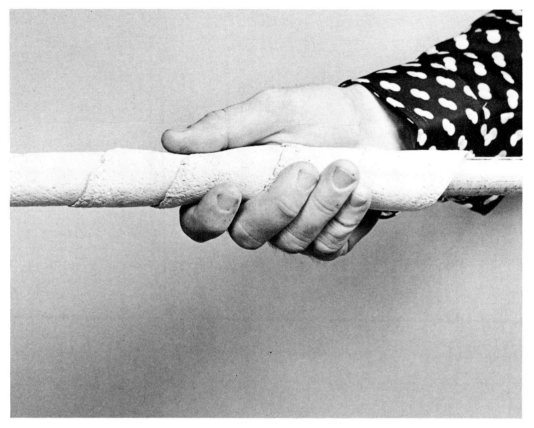

Spiral one length of no-drip tape completely; press the tape together so that it adheres to itself and to the pipe and forms a continuous moisture barrier. The tape is sticky; you don't need any additional adhesive when wrapping it.

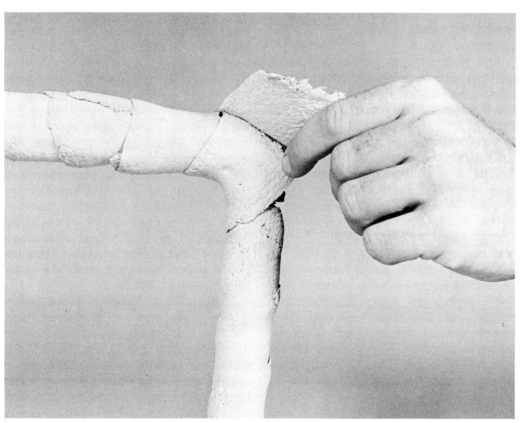

To cover a pipe joint with the no-drip tape wrap the joint so that the tape extends from one wrapped section over the joint to the other wrapped section. The trick is to seal the pipe completely with the tape. You can use scissors to trim away excess tape for a neater job. The tape may be painted, although it is not necessary.

Common Plumbing Problems

An air chamber utilizes a T-fitting in the water supply line. The riser should be about 3 feet long, and one end of it (the top end) should be capped. The pipe should be larger than the diameter of the supply line, if possible; you can buy a reducing connection for this. If possible, install the air chamber at the high point of the plumbing run. Special air chambers are available at plumbing supply outlets.

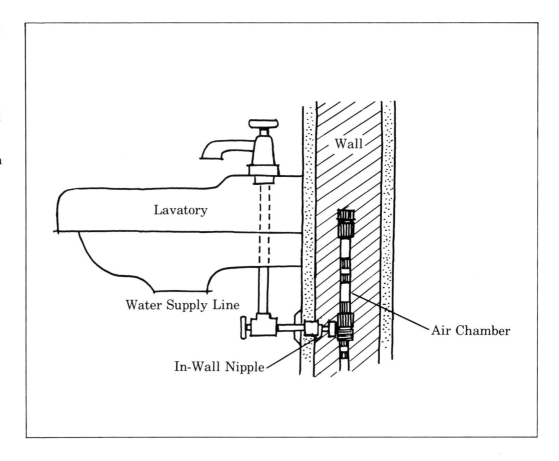

Stopping funny noises in pipes

The water in your home is under quite a bit of pressure. Sometimes when the water flows through the pipes, especially when it goes around corners and is forced through reducing connections, the pipes make noise.

The most common of the various noises is the "water hammer" you hear when you turn off a faucet suddenly. "Water hammer" is caused when water flowing freely through the pipes is suddenly shut off—the water simply slams to a stop.

You can correct "water hammer" by installing air chambers—short, "dead end" pieces of pipe. Air becomes trapped in the chambers and it is compressed by the water flow in the water supply line. When you suddenly turn off a faucet, the air trapped in the air chambers absorbs the shock. (See sketch.)

"Faucet chatter" is that noise you hear when you open a faucet part way and the faucet vibrates. To correct this problem, tighten the washer on the bottom of the faucet stem assembly. If the washer is worn, replace it.

Pipes that squeak are always hot water pipes. These pipes expand and contract when hot water from the water heater runs through them. As the pipe expands, the pipe moves in the hanger brackets. This friction makes a squeaking noise.

To silence the squeak, loosen the hangers slightly. Then stuff a small piece of fiber glass insulation between the hanger brackets and the pipe. Retighten the brackets.

"Angle bang" noise in pipes occurs where pipes make a 90-degree turn. To stop the noise, loosen the hanger brackets, and pad them with fiber glass insulation. If the problem is really bad, you can nail blocks of wood against the pipe joints to stop any pipe movement.

Loose hanger brackets themselves can be the source of the noise. The brackets should be tight, and there should be enough brackets to support the pipe run. If there are not enough brackets, the pipes tend to slap against flooring, studs, and/or joists.

Disposal and hot-water heater drainage

A clogged drain is always a problem with garbage disposals. The problem is usually caused by grease blocking the drain and large chunks of garbage jamming the grinder in the disposal.

Water heaters sometimes spring leaks, or the water won't heat, or there is sediment in the bottom of the tank. Such problems are usually easy to solve, with, perhaps, the exception of a badly leaking hot-water tank which should be replaced.

A jammed grinder and poor drainage are the two major problems with garbage disposals. For proper drainage always use *plenty* of cold water to wash the garbage down; let the water run a long time. If the drainage is completely blocked, remove the drainpipe directly below the sink with a wrench and clean out the pipes with an auger. Do not attempt to open the drain with a chemical cleaner. If the grinder is jammed, make sure there are no knives, spoons, or big bones in the disposal. Then, with the end of a broom handle or rolling pin, try turning the grind ring turntable. (Some manufacturers include a special tool for freeing jammed blades.) When the table is free, run plenty of water into the disposer and turn it on.

If the disposal won't run after you clean it out and the grind ring is free, try pushing the reset button on the side of the disposer housing. If the disposer still won't run, check the main circuit breaker or fuse box. For water leakage, use plumber's putty around the sink flange where the top of the disposal is connected. You will probably have to remove this flange first to get a proper seal.

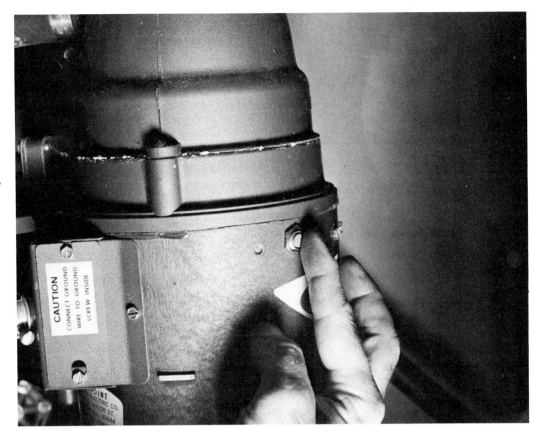

If the hot-water heater is leaking, turn off the water to the tank immediately. The valve is usually located directly above the tank. Next, turn the thermostat to its lowest level of heat, and shut off the gas valve. If the heater is electrically powered, turn off the power at the main service panel, or pull out the outlet plug. If the heater is leaking badly, drain the tank through the valve at the bottom of the tank. Place a bucket under the drain if you don't have a floor drain for the water to run into. Chances are the tank will have to be replaced.

Water too hot or too cold? The thermostat dial is set too high or too low. It should be at the "normal" setting—about 140°. If the water is too hot, the tank may make gurgling sounds. If lowering the thermostat doesn't work, turn off the power to the heater and call in a serviceman. If there is not enough hot water, the problem may be a water heater that is too small for your needs. The standard size today is 40 gallons. Adjusting a thermostat on an electric water heater usually is a job for a serviceman. The thermostat is located under insulation. But check the power source first to make sure the heater is receiving power.

Drain the water heater twice yearly. This will remove any sediment at the bottom of the tank. If the tank is a new one, drain it every 2 months for the first year and every 6 months thereafter. The drain valve looks like a regular hose bib and is located at the bottom of the tank. On some tanks, the handle opens a valve that lets the water run out of the center of the handle. If you have not drained the system in the last 2 years, let it alone. Opening the valve may cause a leak you can't stop.

Common Plumbing Problems 35

Flush tank and toilet problems

A toilet has two parts:
1. The flush tank in which a supply of clean water is stored for the flushing action.
2. The toilet bowl.

Flush tank problems are generally easy to solve, since the flush tank mechanism is simple to repair. Toilet blockage is another story; at best it is messy and takes a lot of patience.

Most flush tank mechanisms work in about the same manner (see sketch). You flip the flushing lever which lifts a ball or rubber flapper inside the tank. Water then flows into the toilet bowl. After the tank is nearly drained, the tank ball or rubber flapper falls back into place.

The float ball operates a water valve. As the water drains out of the tank, the float ball drops with the water level. Just as the water outlet is being closed, the float ball opens the water-supply inlet valve which starts filling the tank with water again through the filler tube.

The water in the tank also goes through the bowl refill tube into an overflow pipe. The purpose of this overflow is to replenish water in the trap seal. When the water level in the tank reaches the top of the overflow pipe, the float shuts off the inlet valve, thus completing the flushing cycle.

The toilet bowl has trap passages through which the water from the flush tank flows.

A wall-hung toilet is held in position by a closet carrier installed on the soil stack. The toilet empties directly into the stack tee through a special fitting.

A floor-mounted toilet is supported by the floor over the end of a branch drain (branches off main soil stack). The toilet is held in place by bolts. (See chapter on special plumbing projects.)

A clogged toilet usually is the result of too much toilet paper in the bowl. The paper becomes wedged in the toilet trap, blocking the flushing action of the bowl. If the paper is not too tightly wedged in the trap, you may be able to dislodge it with a plunger. Work the plunger vigorously up and down. Don't give up too quickly if the obstruction is not cleared instantly. Have the bowl about half-full of water when you use the plunger; water coverage helps provide more suction action.

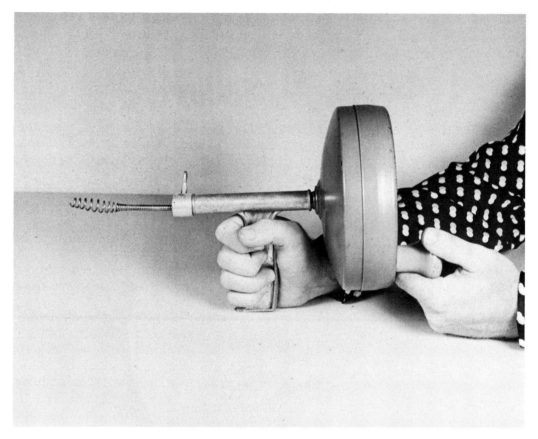

Use an auger if the toilet is really blocked. The auger has a corkscrew device on the end of a rod which will snag the obstruction enabling you to pull it out into the bowl. The pistol-grip handle on the auger lets you hold the auger steady while you rotate the rod drum with the handle.

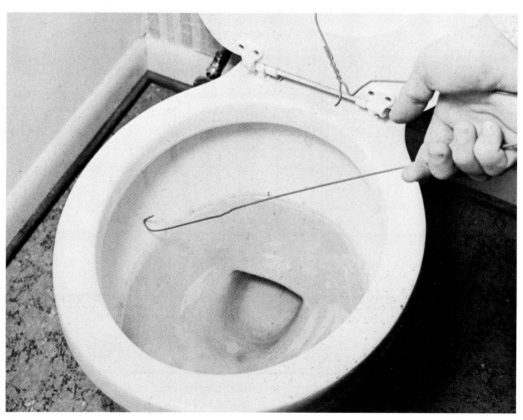

If you don't have an auger, you can make one from a wire clothes hanger. Bend one end of the hanger over with pliers, forming a hook, then work the hooked end into the trap area. Try to snag the obstruction and pull it out into the bowl where it can be removed or broken up and flushed away.

Common Plumbing Problems 37

To use the auger thread it into the trap area as you turn the crank handle. Try to move the auger end into the trap. The trick here is to try to hook the obstruction and pull it out, rather than to push the obstruction down the drain. If the obstruction is pushed down the drainpipe, the blockage may be more difficult to remove.

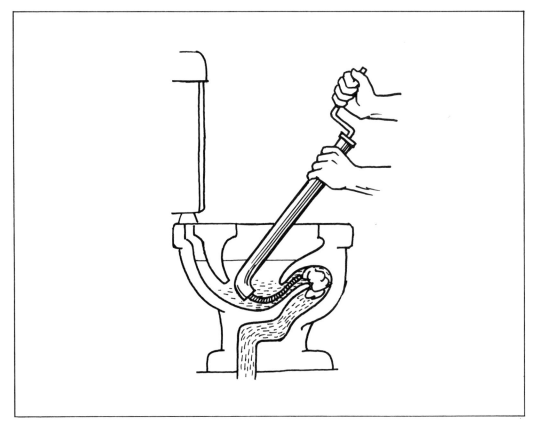

If the obstruction is massive, you may need to remove it with an auger and your hand. You can protect your hand and forearm by wrapping it in a plastic garment bag or a plastic trash bag. Keep the wrapping loose so you have plenty of room to move your fingers and hand. If you are unable to free the obstruction with any of the suggested methods, you will have to remove the toilet from its mounting, invert the toilet, and work at the obstruction from the other end.

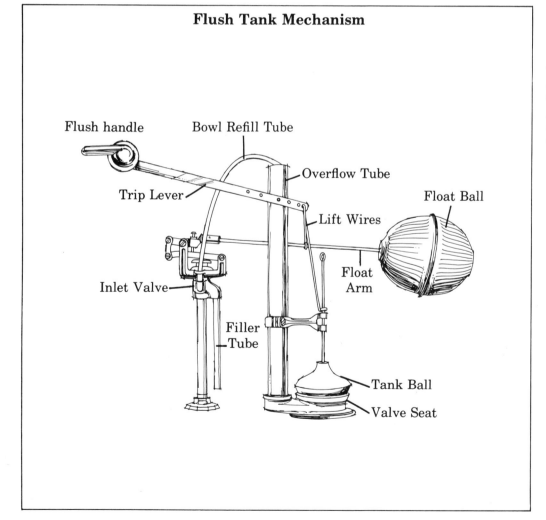

Flush Tank Mechanism

Flush tank problems are caused by a malfunction of this mechanism in the flush tank. Some parts of the assembly may vary slightly in different flush tanks, but the function and repair methods, explained below, are usually the same.

Common Plumbing Problems 39

Adjust the water level in the flush tank. Some inlet valves have a float ball adjustment screw on them as shown. By turning this screw at the top of the valve you can raise or lower the float arm which, in turn, raises or lowers the float ball and the water level.

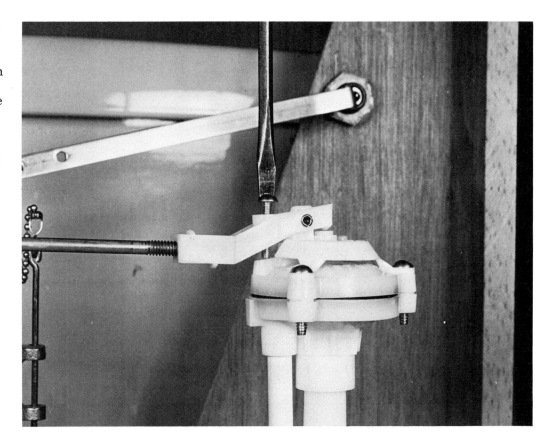

Try bending the float arm if the fixture doesn't have a float ball adjustment screw. If the water is too high, bend the rod down. If the water is too low in the tank, bend the float arm up. The bends should be gentle. Flush the toilet and test the water height in the tank. If the problem still exists, bend the rod a little more. The water should be about 1 inch below the overflow tube in the flush tank.

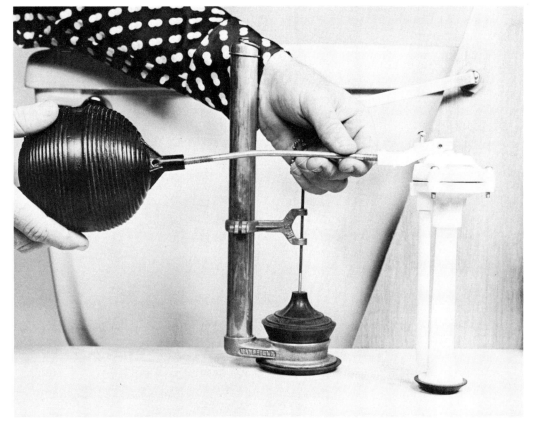

40   Common Plumbing Problems

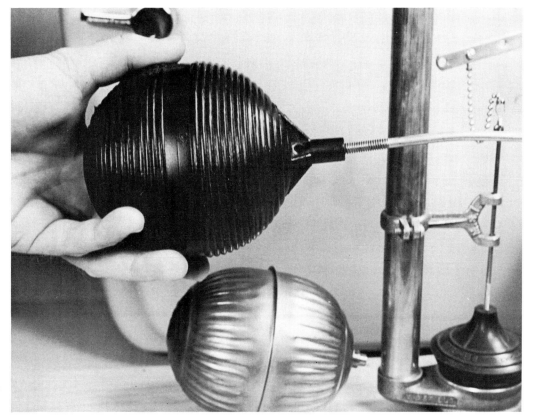

A damaged or leaking float ball can cause low water in the flush tank. Remove the float from the float arm by unscrewing it with your hand. If the float is heavy, chances are it is full of water. The weight of the water in the float prevents it from rising high enough in the tank to fill the tank with water. A water-logged float can also be the cause of a running toilet, since the float can't rise high enough to shut off the water valve in the valve assembly. You can replace the damaged float ball with a new one for about $1.50.

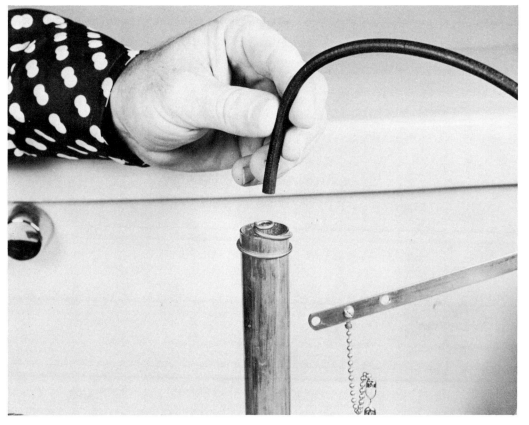

Splashing sounds inside the flush tank. The refill tube may not be discharging directly into the overflow tube. The water, then, flows into the tank causing a splashing noise. A springlike assembly or clip keeps the refill tube in the proper position. If the refill tube is metal, bend the tube slightly so the water runs into the overflow tube. Be careful not to kink the refill tube when you bend it.

Common Plumbing Problems 41

If the flush tank runs constantly, the tank ball lifter may be out of alignment. The lifter probably does not let the ball drop squarely onto the valve seat in the bottom of the tank. If the lifter is bent, replace it with a new one (10¢). The lifting wire must lift the ball straight up during the flushing action.

The guide arm for the lifter wire may be out of alignment, preventing the tank ball from seating properly. This problem also causes the water in the tank to run. Realign the guide arm by turning a set screw which holds the guide to the overflow pipe. Turn the guide with your fingers so it aligns correctly with the lifter wire; then tighten the set screw holding the arm to the overflow pipe.

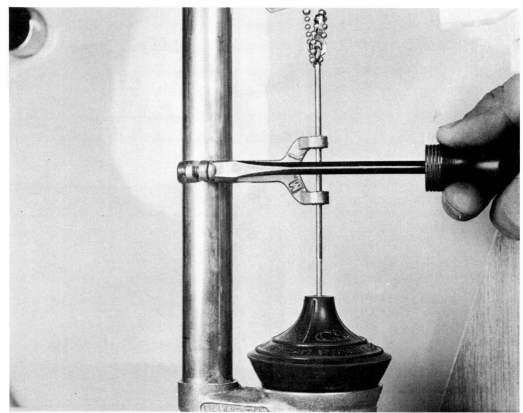

42 *Common Plumbing Problems*

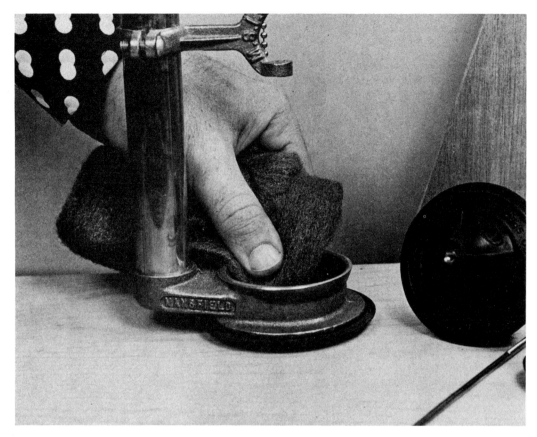

A worn tank ball seat can cause the toilet to run. To correct this, turn off the water, then flush the toilet so the tank is empty. Then scour the valve seat with fine steel wool until the seat is smooth. Test the tank ball in the seat by turning on the water and flushing the toilet. You may have to give the valve several passes with the steel wool until the ball seats properly in the valve.

Toilet difficult to flush? The connection between the flush handle, sleeve, and the trip lever may be rusted or corroded. Or the handle may be loose on the sleeve. You can remedy this by simply tightening the nut that holds the handle to the sleeve inside the flush tank. If this assembly is badly corroded, you should replace it. The cost is not prohibitive—about $3.

Common Plumbing Problems 43

The lifter wire or chain connection to the trip lever has to be fairly tight or the toilet will be difficult to flush. There are usually three hole positions on the trip lever where the connection may be positioned. By trial and error, you can determine the best position for the connection.

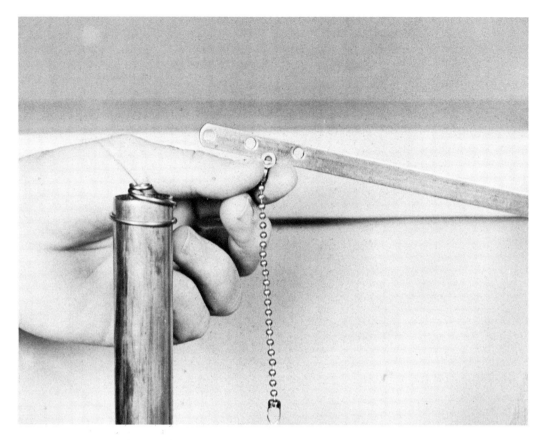

Splashing noises may also be caused by a worn inlet valve. Try flushing the tank and checking to see if the water is leaking around the top of the inlet valve. If so, unscrew the top of the valve and replace the washer or O-ring washer. A new washer often stops the problem. If not, replace the entire inlet valve assembly. A new one costs about $8. To make the switch turn off the water and flush the tank. Have a bucket handy to catch any water when you remove the old valve from the tank. The valve simply screws into position.

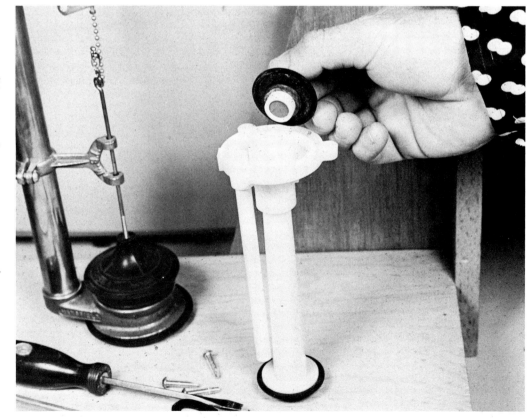

44 Common Plumbing Problems

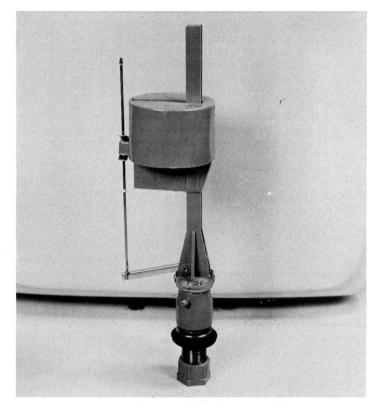

A fluid action ballcock can replace a standard ballcock assembly. This unit has no float ball and works quietly; all component parts are supplied in a kit. To install the fluid action unit, drain the flush tank, unscrew the old unit, and install the new one. By sliding a clip up or down, you can establish the proper water level in the tank. Instructions for installation are furnished by the manufacturer.

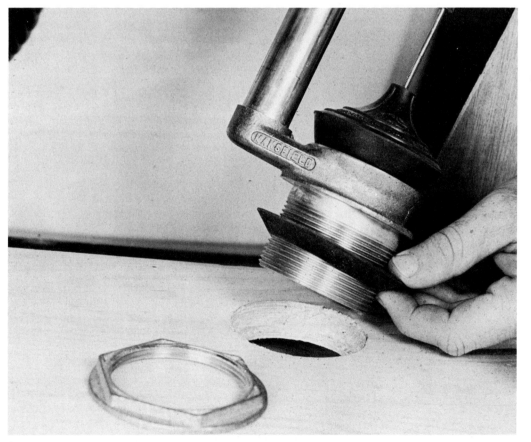

If the tank is leaking at the bottom, chances are that the washers that seal the valve and valve seat are faulty. You can buy new washers to replace the old ones. To install the new washers, turn off the water and empty the flush tank. Carefully unscrew the respective valves, lift them out of the holes, and replace the washers. Go easily with the wrench; too much pressure can crack the flush tank.

Common Plumbing Problems

A cracked flush tank is a common problem—especially when the tank has been in service for a long time. You can buy a replacement tank at plumbing outlets. You should know the name of the tank manufacturer so that a replacement match can be made. Most replacement tanks come with the flush tank mechanism already installed—you simply make the connections.

To replace a flush tank remove the bolts that hold the tank to the toilet bowl. This is the toughest part of the job since the bolts are probably corroded and the nuts difficult to turn. Since the tank is already cracked and worthless, take a hammer and break the tank so you can get at the bolts. But be careful when you do this so that you don't break the toilet bowl. The new tank sits over a rubber gasket, as shown. The new bolts holding the tank in position go down through the inside of the tank and through the back of the toilet bowl. Tighten each bolt the same number of turns. This distributes the stress on the tank evenly and prevents cracking.

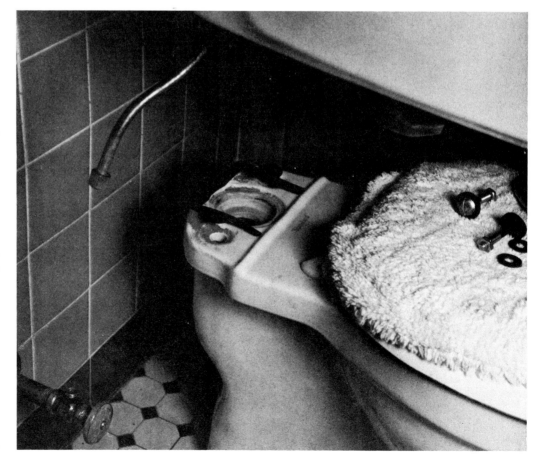

Hook up the water to the flush tank. The water supply line is usually flexible so that you can make gentle bends in it to fit the fixture above. The supply line has compression fittings on each end; don't overtighten the compression nuts. When the tank is in position and connected, turn on the water and watch for leaks. If you spot a leak, you will have to turn off the water, empty the tank, and tighten the bolts some more.

Galvanized Steel and Cast-Iron Pipes and Fittings

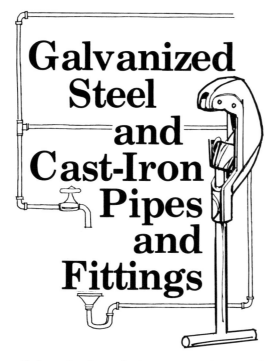

Galvanized steel pipe or cast-iron and threaded steel pipes are used for stacks, drains and sewers, and drains buried underground. They are also used for water supply runs.

Galvanized steel pipe is a quality product. As a rule, you can purchase steel pipe that is threaded at both ends; it is fairly easy for a home handyman to assemble since the joints are simply screwed together. The pipe is available in a wide range of lengths and sizes which means you don't have to buy—or rent—threading equipment except for specialized jobs.

Most drainage systems are designed using cast-iron pipe and fittings for all main stacks, any toilet branch drain, the house drain, and for any pipe branch that is buried either underground or in concrete. Cast-iron pipe and fittings are available in 2-, 3-, and 4-inch sizes; pipe, in 5- and 10-foot lengths.

Galvanized steel pipe and fittings are used for all other parts of the drainage system, including secondary soil stacks, vent lines joining a stack, and branch drains other than those listed under cast-iron pipes. Galvanized steel pipe and fittings are available in $1^{1}/_{2}$- to 2-inch sizes; pipe, in 10-foot lengths.

"Sanitary" drainage fittings, designed especially to avoid any solid accumulation in the pipes, are used in all drainage lines that carry waste water. All cast-iron fittings are of the sanitary type.

"Standard" (non-sanitary) and "regular" (non-sanitary) threaded fittings are used in vent fittings where only gases are carried. Galvanized steel fittings are available in both the sanitary and standard types.

Regular galvanized steel pipe and fittings are mainly used in water-supply systems, particularly in areas where pipes may be subject to damage; galvanized pipes resist shocks and blows. These pipes are available in standard sizes of $^{1}/_{2}$, $^{3}/_{4}$, and 1 inch and in pipe lengths up to 21 feet. Pipe lengths are available already threaded, and the fittings are similar to those used in drainage systems described above.

Depending on the building codes in your area, you may be able to use plastic pipe or fiber pipe for drainage systems. You also may be able to use plastic pipe for cold water systems. This is discussed in a later chapter. Copper and clay pipe may be used for drainage as well as water systems; this is also discussed later.

Although *Easy Home Plumbing* concerns itself mainly with plumbing emergencies and repairs, you should check out the plumbing codes in your area before you attempt any project or repair.

Some cities, for example, require work permits before any plumbing work is started. When the work is completed, the codes specify that the work must be inspected and approved. In addition, some cities require that only licensed plumbers do the work. Still in other cities, the codes will let you do the work, but the final job has to pass an inspection. Codes often apply to repair work as well as any new construction work.

Distance Pipe Goes into Standard Fittings	
Size of pipe	Distance into fittings
½ inch	½ inch
¾ inch	½ inch
1 inch	9/16 inch
1¼ inch	⅝ inch
1½ inch	⅝ inch
2 inches	11/16 inch

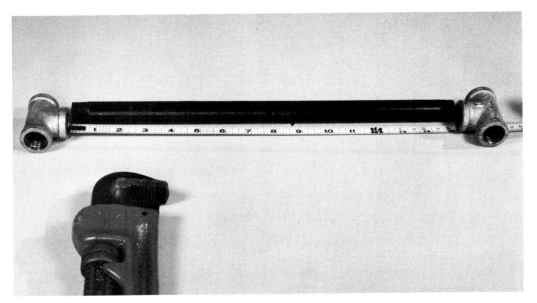

Pipe length is measured from the face edge of one fitting to the face edge of the other fitting. The length of the pipe that will go into the fittings at both ends is added to the basic length of the pipe (see chart).

Cutting and threading steel pipe

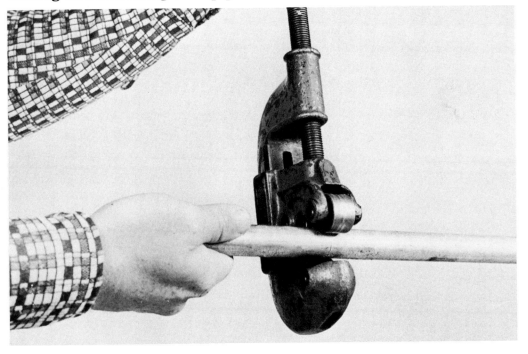

Cut galvanized steel pipe with a hacksaw or steel pipe cutter. The cut must be square so that the pipe can be threaded. To use a pipe cutter slip the cutter over the pipe with the wheel matching the cut-off mark. Tighten the wheel and rollers. Rotate the cutter around the pipe a couple of turns; then tighten the wheel again. Use thread-cutting oil on the cutter wheel as you go. Support the cut-off end near the end of the cut to prevent the end from sagging. If you use a hacksaw to cut the pipe, remember that the saw cuts on the forward stroke. Keep the strokes uniform.

Pipes and Fittings

Right:
Remove excess flared metal from inside the cut pipe with a pipe reamer. Use a metal file to remove excess metal outside the pipe and at the cut end.

Far right:
To cut external (male) threads use a solid die, shown here, or a split die. Dies fit into a threadcutter assembly—one or two long handles for applying leverage to the dies—which you rotate clockwise until the die "bites" into the pipe cutting threads.

Start the correct size die on the end of the pipe at right angles. Turn the cutter clockwise until the die engages the metal of the pipe. Apply cutting oil generously to the pipe and die. The oil helps reduce friction and heat on the pipe. As you cut the threads, back the die off from time to time to remove metal chips produced by the cutting action. When the end of the pipe is flush with the surface of the die, the threads are the proper length. Test the threads by screwing on a fitting as far as you can by hand. Then, with a pipe wrench, continue turning the fitting so that two threads are left showing on the pipe.

50   *Pipes and Fittings*

Assembling steel pipe

Pipe joint compound, or a plastic-type tape made for this purpose, is always used when assembling galvanized steel pipe runs. Apply the compound to external (male) threads sparingly. Never apply compound to internal (female) threads. Keep the compound out of the inside of the pipe.

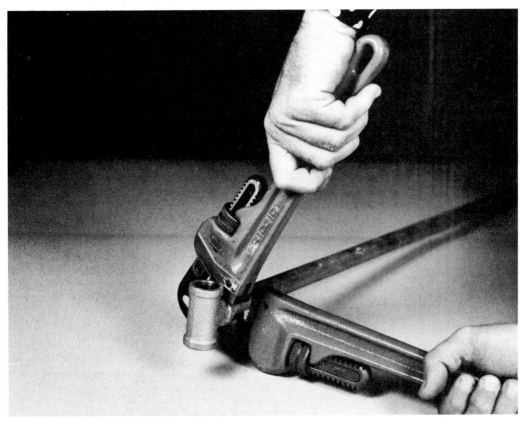

Double-wrench pipe assembly. When connecting pipe runs where couplings or fittings are concerned, put one wrench on the fitting and another wrench on the pipe. The jaws of the wrenches should face in opposite directions. This prevents strain on the pipe and fittings. Pipe wrenches are so powerful they can crush the walls of pipe and fittings, so be careful; do not overtighten fittings.

Pipes and Fittings 51

Union fittings, shown here, are a must in fabricating or repairing a galvanized steel pipe run. The unions must be installed where you cut through a pipe or where you add a pipe. Standard fittings turn one way; union fittings turn in opposite directions so the plumbing run can be completed. Always match thread size to the prethreaded pipe when purchasing pipe and fittings. Drain fittings (sanitary) are smooth inside. Standard fittings (non-sanitary) are used in vent systems only. Plastic, bronze, copper, and brass fittings are for water-supply lines only.

To remove a section of leaky pipe turn off the water; then cut the pipe between two fittings with a hacksaw. Make the cut at a slight angle. Put one pipe wrench on a fitting and, with another pipe wrench on the pipe, remove the pipe from the fitting. Do the same at the other end of the cut pipe—one wrench on the fitting with another wrench on the pipe. This protects the fitting from damage.

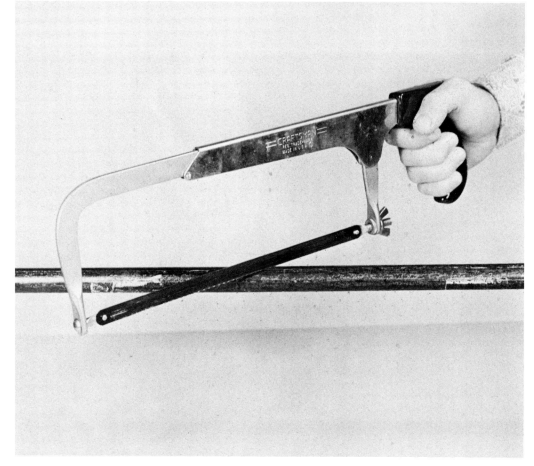

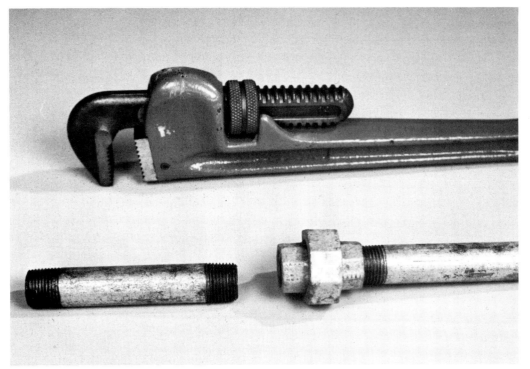

Run one end of the new pipe into the fitting. At the opposite end of the new pipe install a union fitting as shown. Then run a second piece of pipe into the union and the opposite fitting. Use pipe joint compound on the male threads. Also use two pipe wrenches—one on the fittings and union and the other on the pipe. Using two lengths of pipe and a union is easier than trying to rethread pipe and couple it for replacement.

A union fitting has three sections. One section is threaded onto the pipe as shown here. Start the fitting with your fingers. When you are sure that the threads are tracking properly, tighten the joint with a pipe wrench. You also should have a second pipe wrench on the pipe.

Pipes and Fittings 53

A second, flange-type section of the union fitting is threaded onto the opposite piece of pipe. The union housing then links the two pipes together and makes a watertight seal between the pipes.

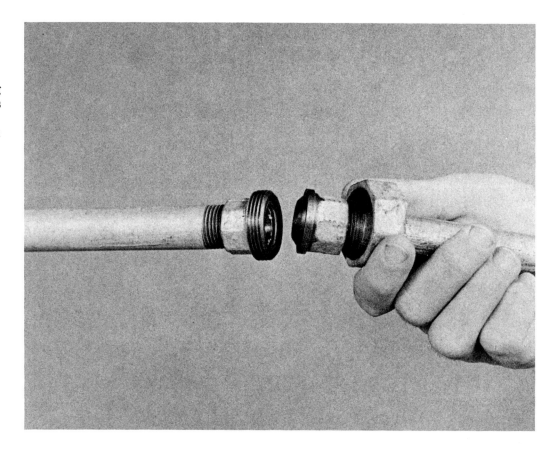

To add a new pipe run to an old run, you don't always have to dismantle the plumbing. You can buy an adapter called a "saddle tee"—shown here—that clamps around the existing pipe run. The adapter has a gasket that seals the joint. To use the adapter turn off the water; then drill a hole through the existing pipe. Insert the drill bit into the opening in the adapter for the drill. Hook on the new pipe.

54 *Pipes and Fittings*

Hub-and-Spigot Cast-Iron Pipe and Fittings

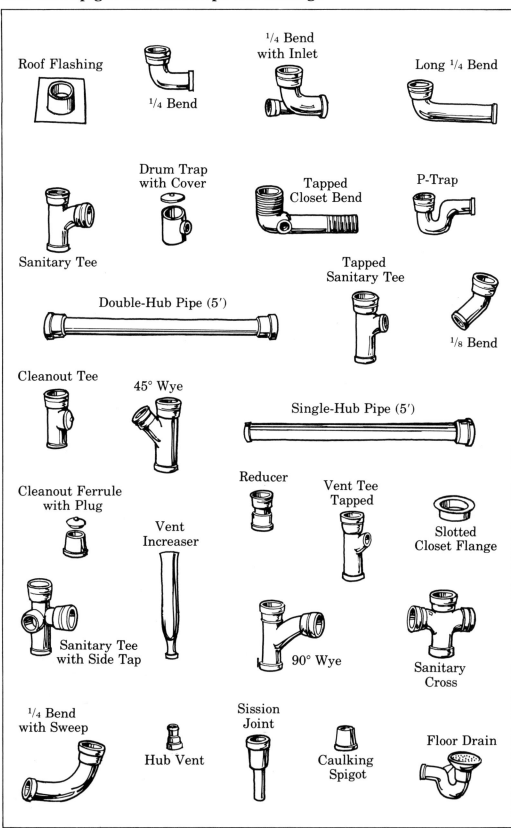

Hub-and-spigot cast-iron pipe and fittings are connected by inserting a spigot end into a hub end and then caulking the joint. (The words "tap" and "tapped" used in identifying some of the fittings mean that the particular fitting has a threaded opening for a steel pipe connection.)

No-Hub Cast-Iron Pipe and Fittings

No-hub cast-iron pipe and fittings are easily connected by the use of special sleeve couplings.

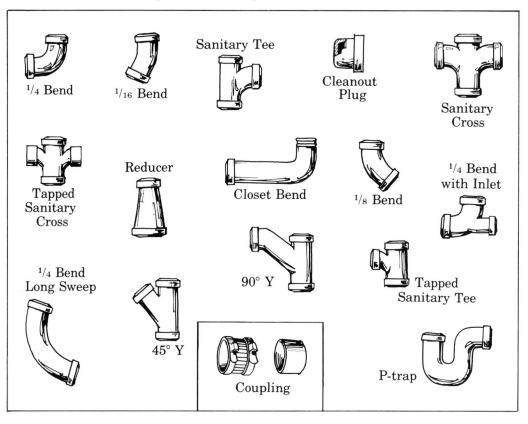

Standard Steel Pipe Fittings

Steel pipe and fittings are connected by screwing an externally (male) threaded end into an internally (female) threaded fitting.

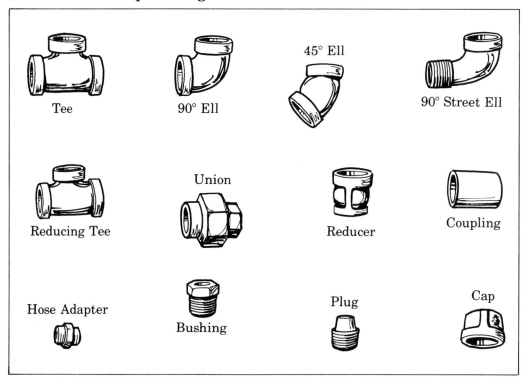

Working with drainage pipe

Cast-iron pipe is probably the most popular material for construction of soil and waste stacks. The pipe has a bell-shaped "hub" at one end; the other end is rigid and termed a "spigot." The spigot end fits into the hub.

You can also purchase cast-iron pipe with no-hub joints. Such pipe is easier to work with than hub-and-spigot pipe since neoprene gaskets are used at the joints instead of lead, caulking, and oakum. Too, the gaskets permit slight variations in alignment, so you don't have to align each piece perfectly. There also is less wasted pipe length with no-hub pipe.

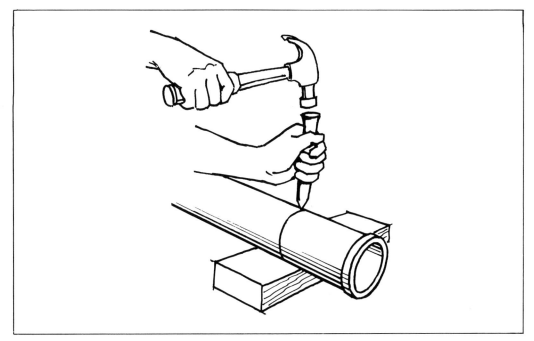

To cut cast-iron pipe mark a cutting line on the pipe. Use a hacksaw and make a groove at the mark about 1/16 inch deep, turning the pipe slowly to complete the groove all around the pipe. Lay the pipe on a 2- by 6-inch board. Use a cold chisel and hammer to tap around the groove. Pound lightly, going completely around the pipe. Continue this process until the pipe breaks.

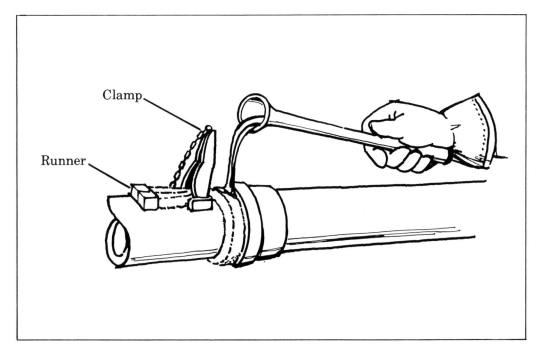

Make horizontal joints in a vertical position, if possible. This can be done with short lengths of cast-iron pipe. If the joint must be made in the horizontal position, you will have to have a joint runner to prevent molten lead from spilling out of the joint as it is poured. Clamp the pliable runner around the spigot end of the pipe and up against the hub. This seals the pipe so that you can fill the pipe with lead.

To assemble a vertical joint with cast-iron pipe, first assemble the pipe and pack the hub with oakum. Oakum is available in strands about ½ inch thick. The depth of the oakum should be about 1 inch from the top of the hub. Pack it firmly in the hub using a cold chisel.

Fill the hub with molten lead. It takes about 1½ pounds of lead to fill a 2-inch pipe joint. The lead should be just hot enough to pour but not hot enough to burn the oakum.

Caulk the hub joint when the lead has cooled. Lead tends to shrink when it is cool; thus the necessity for the caulking. You will need a plumber's caulking iron and hammer for this job. The caulk should be uniform and tightly packed. Don't hammer too hard, however; you could crack the joint.

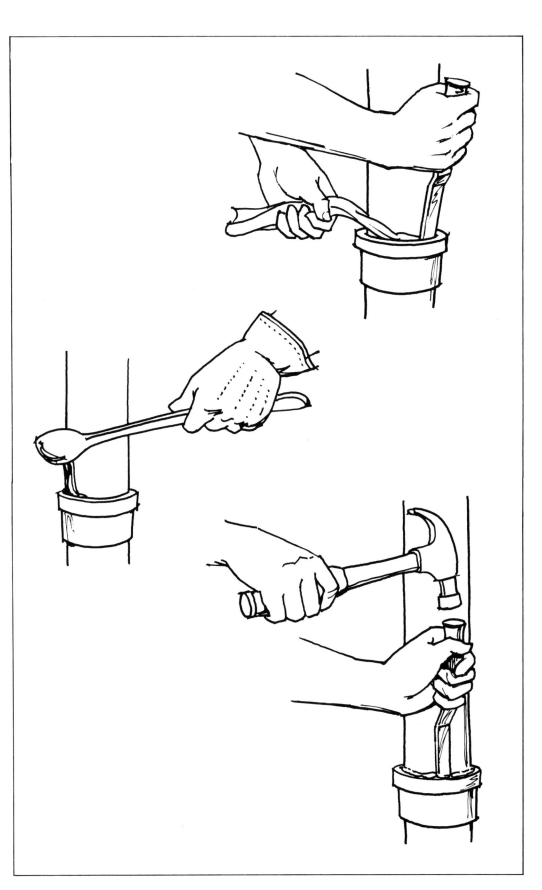

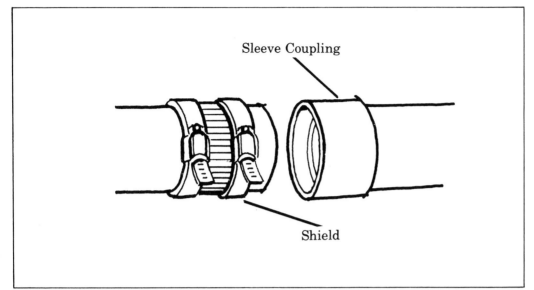

No-hub pipe joints are made with a special sleeve coupling. Slip one end of the sleeve over the pipe as far as it will go. Put the stainless steel shield with clamps around the other pipe end.

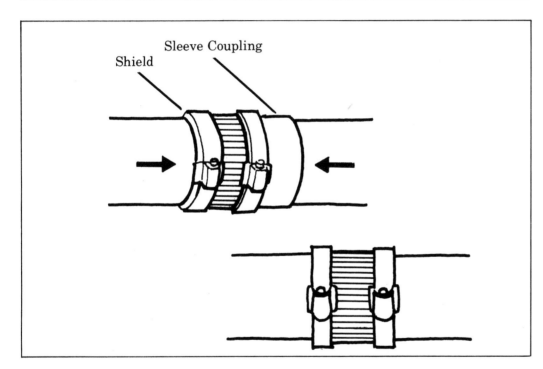

Push the second pipe end into the sleeve as far as it will go; the pipe ends should be tightly butted together. There is a molded "shoulder" inside the coupling which should be firmly "sandwiched" between the two pipe ends. Finally, slide the metal shield to the center of the sleeve coupling and tighten the two clamps with a torque wrench set to release 60 pounds.

Fittings are joined in exactly the same manner as straight joints. Special adapters are available for fitting no-hub pipe to other types and sizes of pipe.

Troubleshooting septic systems

Because of the work involved, you probably will not want to dig trenches and lay pipes for a septic system.

Breakdowns, too, should be left to a professional. Most of the component parts of the system are underground in the form of a tank, well, and drain field.

Your biggest problem with a septic system probably will be bacteriological action in a tank that is flooded with water. Flooding will cause a backup in the drain piping. You will notice this in your lawn: the grass will be greener above the drain lines than elsewhere in the yard.

Your septic system should be inspected by a professional about every two years at which time he should pump out the sludge accumulation in the tank.

Pipes and Fittings 59

Disposal field for septic system should be on flat land—if possible—and as far as possible from the house and other occupied buildings. Have a professional service your septic system about every two years or as necessary. His expertise is well worth the service call which usually runs about $30.

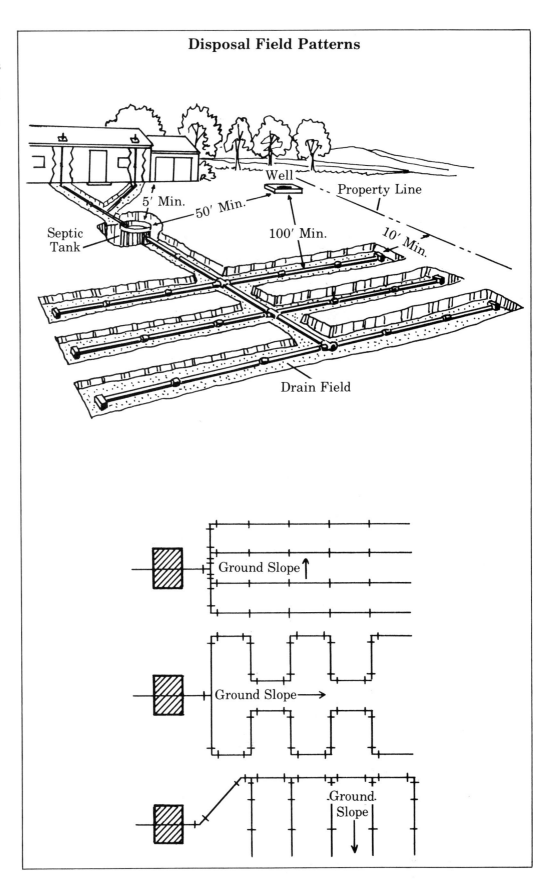

Disposal Field Patterns

60 *Pipes and Fittings*

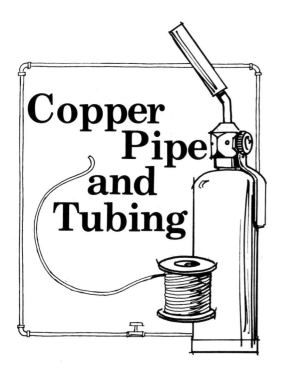

Copper Pipe and Tubing

The chief advantage to copper pipe and tubing is that the material is lightweight and need not be threaded for joints. The pipe and tubing are manufactured in both rigid and flexible lengths, both of which resist corrosion and are quite durable.

Copper pipe is available in three weights:
1. Type K—thick-walled
2. Type L—medium-walled
3. Type M—thin-walled

Type K copper tubing is usually permitted for use underground. Type L tubing may be used for any residential plumbing. Type M tubing is for waste or drainage lines, but the problem with type M tubing is that it takes a lot of heat to join the material together.

Rigid copper pipe is probably easier to work with than its flexible cousin because the joints are always straight. But flexible copper tubing is better used for vertical installations where you must feed the pipe through the wall or ceiling.

Like any other plumbing project, check local plumbing codes before using copper pipe or tubing.

For most plumbing jobs you will be working with $3/8$-, $1/2$-, and $3/4$-inch copper pipe, depending, of course, upon codes. The outside diameter of this material runs $1/8$ inch larger than its nominal size.

Flexible copper tubing is manufactured in Type K and Type L. You probably will want to choose Type L since it may be used for almost any plumbing project you plan. Type K is manufactured especially for heavy-duty use.

Copper pipe and tubing is easily assembled "dry" before making any solid, permanent connections. This feature can be a great advantage to you, saving time and money in wasted materials—especially if you have made errors in your layout calculations. However, measure the project carefully before buying any materials or tools. Determine what type of joint will be needed for the job—sweat-soldered joints, flared joints, or compression joints. The different types of joints are shown in this chapter.

Tools for working with copper pipe and tubing are fairly inexpensive, but the type of tools you choose should depend on the joining techniques your job requires and the type of pipe or tubing. For example, if the job dictates sweat-soldered joints, you need not purchase a flaring tool.

For both copper pipe and tubing you will need a hacksaw or a tube cutter. For most jobs, a tube cutter is the better choice. Yet the price of the hacksaw is so nominal, it will probably be beneficial in the long run to buy both the tube cutter and the hacksaw.

You will also need a propane torch if the joints are to be sweat-soldered, as opposed to flared or compression fittings. If the joints are to be flared, you will need a flaring block and flaring tool. There is a flaring tool that may be driven with a hammer, but this technique can be risky since you must hit the tool squarely with the hammer. Any mistakes can cause the pipes to leak.

If the joints are to be compression fittings, you will need a set of adjustable wrenches. Flared joints and compression joints are similar in assembly. However, compression joints are slightly more expensive to make than flared joints. On the other hand, compression joints are easy to assemble and disassemble.

Other necessary materials include steel wool, soldering flux, and a spring tube bender if the tubing will need to be bent. If only slight bends are required in your project, you can make the bends by pulling the tubing down over your knee. Be careful not to kink or bend the tubing at right angles. A good rule to follow: the larger the diameter of the tubing, the less you are able to bend it. Large diameter tubing tends to kink.

Copper Pipe and Tubing Sizes

Nominal size of pipe in inches	Outside diameter in inches	Inside diameter in inches		
		K	L	M
3/8	.500	.402	.430	.450
1/2	.625	.527	.545	.569
3/4	.875	.745	.785	.811
1	1.125	.995	1.025	1.055
1 1/4	1.375	1.245	1.265	1.291
1 1/2	1.625	1.481	1.505	1.527

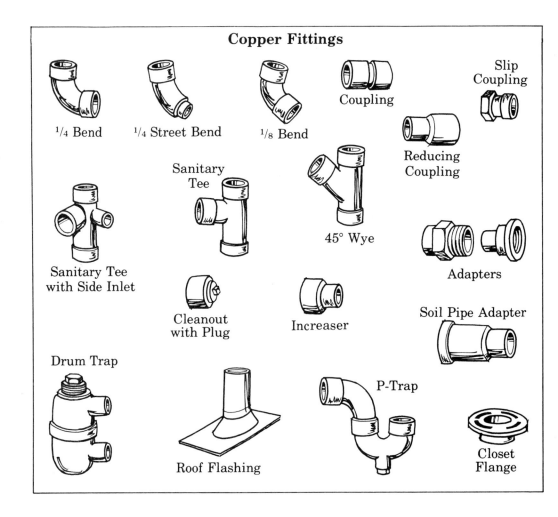

62 *Copper Pipe and Tubing*

Sweat-soldered copper fittings

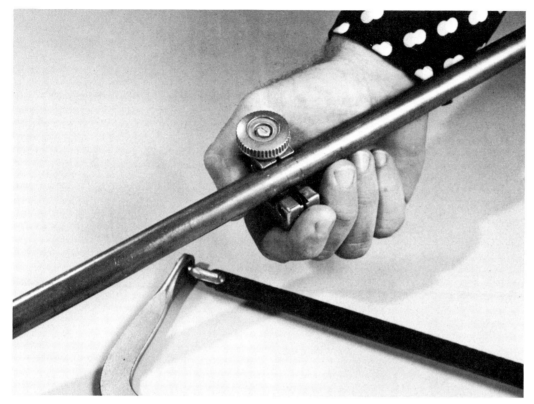

Cut copper tubing with a special tube cutter for sizes up to $1^{1}/_{2}$ inches. Use a fine-toothed hacksaw for larger sizes. The tube cutter, which is inexpensive, has a cutting wheel. Rotate the wheel around the tubing, as shown. After every two circles, tighten the thumbscrew, which applies more pressure to the cutting wheel. If you use a hacksaw to make the tubing cuts on larger size pipes, make sure the saw is square on the tubing.

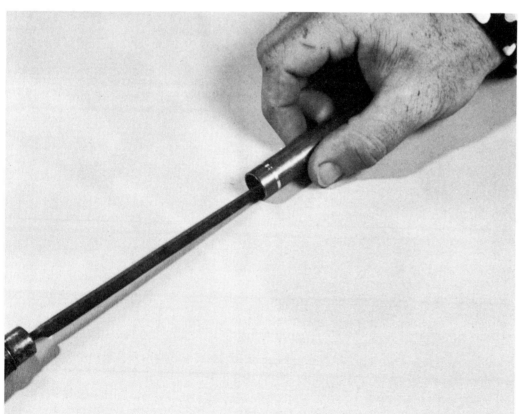

With a file smooth rough edges of the tubing where you cut it. File the inside of the pipe as well as the outside. All burrs must be removed so that the pipe will fit snugly into the hub. The inside of the fitting hub must also be unmarred and perfectly clean.

Copper Pipe and Tubing 63

Shine the ends of the tubing and inside the fitting hub with fine steel wool until they sparkle. The steel wool removes all grease from the copper so that the solder and flux can form a good bond. Any dirt, moisture, or oil from your hands will prevent a good weld. Check the end of the pipe where it goes into the fitting; the end must be square and free from burrs and nicks. Prefit the connections, making sure they fit tightly. There should be no sloppy fits.

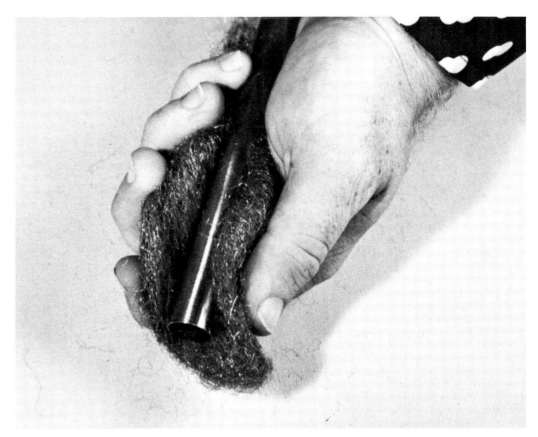

Apply soldering flux evenly to the end of the tubing after it is properly cleaned. Apply the flux to the inside wall of the fittings as well. Flux is used to clean, to wet the surfaces so that solder will flow evenly, and to prevent the heat from a propane torch from oxidizing the copper. Use only noncorrosive flux; do not use acid flux on copper.

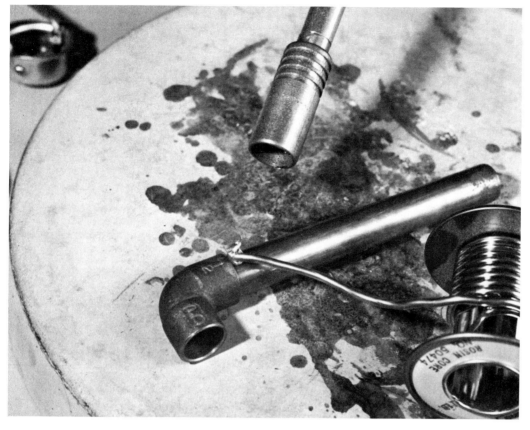

Use solid-core wire solder, never soft-core solder. The best solder to use for copper joints is "50-50" wire solder—50 percent lead, 50 percent tin. It flows at 250° F. and is slow to harden. To assemble copper joints, first insert one end of the pipe into the fitting. Heat the joint with a propane torch, using just the tip of the flame (the tip is the hottest point). When the fitting reaches soldering temperature (test it by touching with solder), apply the solder to the joint. The solder will form a tiny fillet around the shoulder of the fitting. Keep the flame evenly distributed on the fitting and pipe, and do not overheat the metal.

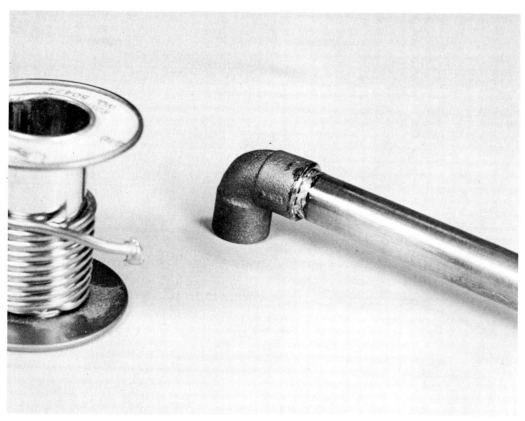

The completed joint should be full of solder—no gaps around the shoulder of the joint. When you make any repairs on copper tubing water systems, be sure to drain the pipes after you turn off the water. Copper tubing *must* be dry when you work with it. If it is not, the solder will not adhere properly to the metal. If you have to break a joint for repairs, heat the joint and pull it apart—after you turn off the water. To reassemble the joint with new pipe, or old, shine the ends of the tubing, apply flux, and solder the joint.

You can bend copper tubing with a special springlike tube bender. Simply insert the tubing into the bender, turning the tubing clockwise as you push it in. The tubing should protrude beyond the point where you want the bend. Bend the tubing over your knee. To remove the bender, twist it off the tubing.

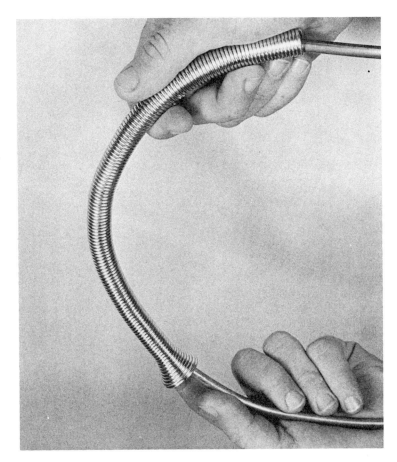

Flared and compression copper fittings

Flexible copper tubing may be assembled with compression joints and fittings. Or you may solder it to the joints as described earlier. Slide compression fittings onto the tubing; then assemble the joint with a screw fitting as shown here. The advantage of using compression fittings (and flare joints) is the flexibility of working with the copper and the fact that the joint may be easily disassembled at any time with minimum work. Compression and flare fittings, however, are more costly than regular soldered joints.

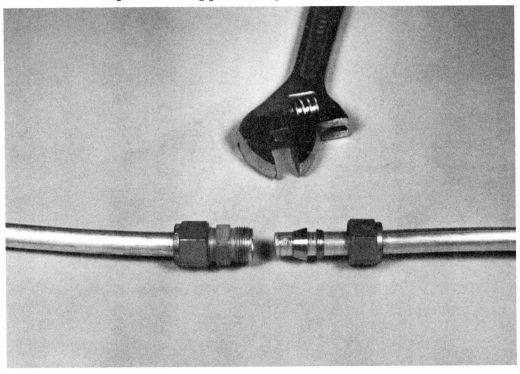

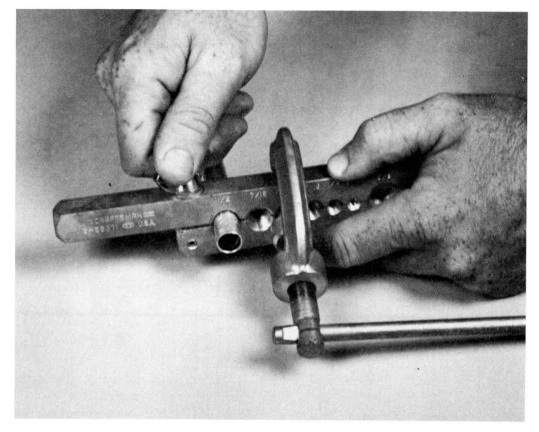

The easiest way to make a flared joint with copper tubing is to use a die and a flaring tool. You can use a less expensive hand-held flaring tool teamed with a hammer to make the flare. To use the die slide the flare nut on the tubing; then lock the tubing in the die as shown. The end of the tubing should protrude slightly above the face of the die.

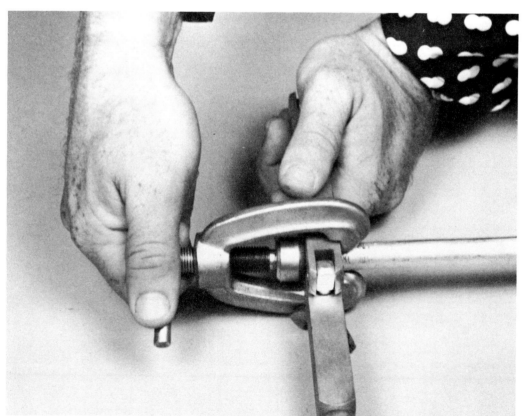

Clamp the flaring tool over the die. Twist the tapered point into the tubing. You shouldn't have to use a lot of pressure. If you use a hammer-driven flaring tool, insert the tapered end of the tool into the tubing, and strike the other end of it lightly with a hammer until you form the desired flare. Make sure you strike the flaring tool squarely.

Copper Pipe and Tubing 67

Flared tubing looks like this when it is removed from the die. Slide the flare nut down on the tubing, and check the fit. If the flare is not properly seated in the flaring nut, lock the tubing back in the die, and apply more flaring pressure. When you make the connection, use two adjustable wrenches on the flare nuts, applying equal pressure. If the joint leaks when you turn on the water, you will have to reflare the tubing. Here, cut off the old flare and start anew.

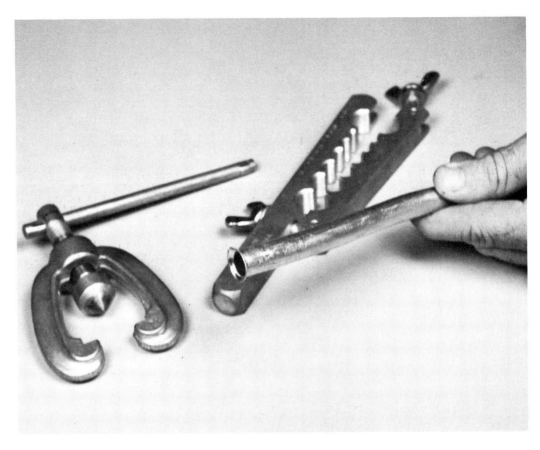

Compression fittings are ideal for sink and lavatory connections and some flush tank connections. The joints go together very quickly and are easy to disassemble in the event of plumbing problems. You can use compression fittings on aluminum and chrome tubing as well as copper tubing. The fittings should match the metal, however.

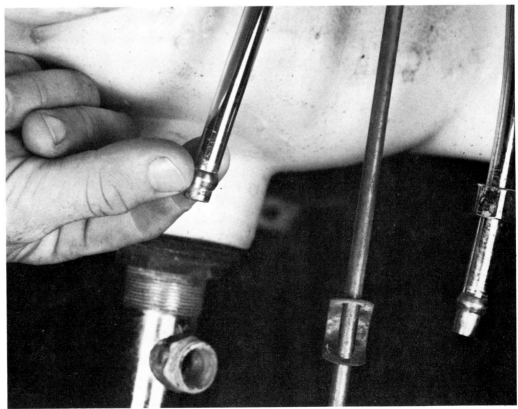

Plastic Pipe and Tubing

A relative newcomer on the building scene, rigid plastic pipe is an easy product to work with. It is lightweight, and the joints fit together like a tinker toy set. If you make a mistake in measurement, the pipe is quick and easy to disassemble.

There are three types of plastic pipe used in home plumbing:
1. PVC (polyvinyl chloride)
2. CPVY (chlorinated polyvinyl chloride)
3. ABS (acrylo-mitrile butadiene-styrene)

All three types of rigid plastic pipe can be used in the cold water-supply system. Only CPVC pipe can be used in the hot water-supply system and your home's drainage system.

CPVC pipe is usually available in 1/2- and 3/4-inch sizes. For hot and cold water, the 1/2-inch size is generally specified. The 3/4-inch material is used for main distribution lines. You also can buy larger sizes of rigid plastic pipe for main stacks, toilet branch drains, and house drains (3-inch material). Two-inch plastic pipe may be used for secondary stacks and some branch drains.

Plastic pipe has several advantages: the material is corrosion-proof. It will not become rusty from acid, alkali, or corrosive water, and it is not affected by electrolytic action. The pipe is also a natural insulator; hot water stays hotter in the runs and cold water stays colder.

Pipe ratings are stamped on the pipe in pounds per square inch (psi). The rating should match the water pressure in your city. The plumbing outlet where you buy the pipe can supply this information.

Flexible plastic tubing is made from polyethylene. It is usually black in color and may be used for such outside projects as underground sprinkler systems, wells, and supply lines to gardens. The polyethylene tubing, generally sold in 100-foot coils, should never be used for hot water runs.

The flexible tubing is usually available in three pressure categories: the best is rated to withstand up to 125 psi. The medium-quality flexible plastic tubing is rated to take up to 100 psi; the utility grade is used only for low pressure installations such as sprinkler systems.

Some local codes restrict the use of plastic pipe and tubing to lawn sprinkler systems. Use of the material within or behind walls is generally forbidden. Always check the plumbing codes in your area before using this material for new or remodeled plumbing runs.

Recommendations for connecting plastic pipe to copper tubing and iron pipe are given below. When you purchase plastic pipe, be sure to ask for any available manufacturers' literature on specific procedures in making the connections. The procedures here are general. However, they should apply in most installations.

When connecting plastic pipe to copper tubing, you can break a 3/4- or 1/2-inch copper line almost anywhere in the system. Then add a take-off tee to the line, using a copper tubing adapter.

If you are connecting into an iron pipe system, try to add the plastic pipe as closely as possible to a union joint in the iron pipe. This will save plenty of time, since you can work from the union back to an existing elbow. The elbow may then be replaced with a new iron pipe tee, and the plastic pipe connection can be made at this point.

If you have to cut a straight section of existing iron pipe, try to select a short section. You will have to replace the old iron pipe with a new section of plastic pipe. Thread the adapters onto the iron fittings, and use a tee in the center of the new pipe as a take-off for the new plastic line.

At this point you should measure the length of pipe you need. Measure along the exact line the new pipe is to follow. If you need more than one length of pipe for the run, make sure you note how many different couplings you will need for the project.

(All types of plastic pipe may be used in these recommendations except that you must use CPVC pipe in the hot water supply system and in the drainage system as noted above.)

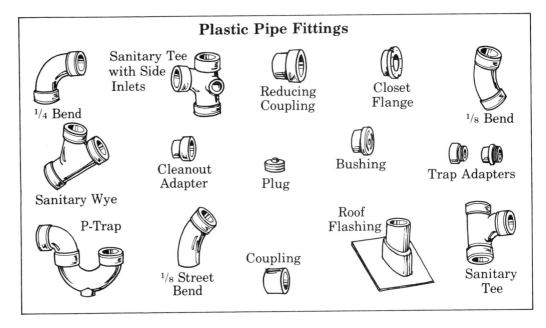

Cutting and assembling

To cut rigid plastic pipe, use a hacksaw with a light forward cutting pressure. Keep the hacksaw square to the pipe so the fittings will seat properly. When fabricating long runs with plastic pipe, allow for 1/2-inch length of material for the fittings; use the longest length pipe you can (10 feet is standard). Example: the run is 8 feet long with two couplings. The length then should be 8 feet 1 inch. This way, you will avoid unnecessary coupling joints. Rigid plastic pipe in long runs must be supported with hangers.

70 *Plastic Pipe and Tubing*

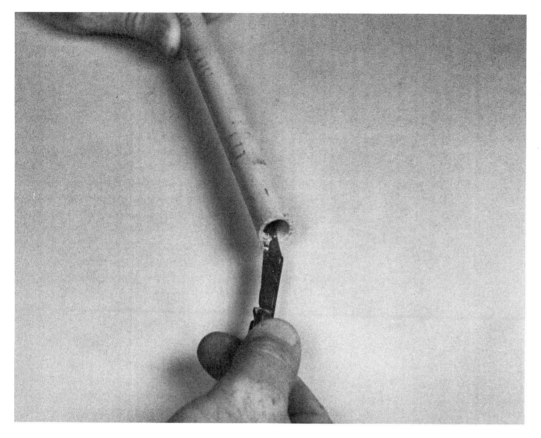

Remove any burrs from the inside of the pipe with a knife. Simply run the cutting edge of the knife around the inside edge of the pipe.

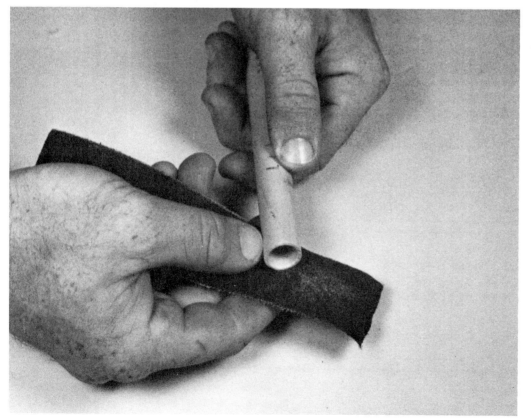

Sand the ends of the plastic pipe with fine grit abrasive. All burrs and nicks should be removed before the fittings are cemented in position. When you have finished cutting, trimming, and sanding, assemble the pipe run "dry." Do not cement the fittings until you are certain that all elements fit together perfectly.

Plastic Pipe and Tubing 71

Apply the cement to the pipe. A small, artist's brush does a good job; do not stick the pipe end into the cement. Be sure to use cement recommended only for the type of pipe you are using. Twist the fitting or coupling on the pipe, and center it in the position you want. The cement should form a tiny bead at the shoulder of the fitting or coupling. If it does not, remove the fitting and recement it. Once the cement is dry— and it dries quickly— you cannot remove the fitting without cutting the pipe.

Flexible plastic tubing is also cut with a hacksaw with light forward pressure. Since the material comes in coils, it is best to unroll the coil and let the plastic material soften and straighten before you work with it. Clean away any burrs left by the hacksaw with a razor knife.

Flexible tubing is fastened to fittings with clamps instead of cement. Use stainless steel worm clamps, particularly when joining the tubing to non-threaded fittings. There are many different types of fittings available: elbows, straight connectors, tees, etc. The fitting should fit the inside diameter of the tubing.

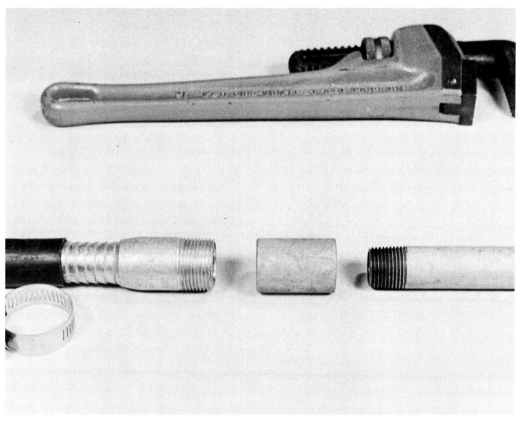

Flexible plastic pipe may be joined to steel pipe with a steel fitting and coupling as shown. It may also be joined to itself with polystyrene fittings and couplings. If you have to disassemble a joint, pour hot water on the plastic pipe. This will soften the plastic and make it easier to break the seal by pulling the joint apart.

Plastic Pipe and Tubing 73

Plumbing Projects

With, perhaps, the exception of changing or modernizing plumbing fixtures, plumbing remodeling or additions require a floor plan of the project and a plumbing plan of the project.

Floor and space planning are not dealt with in this book; but such information is readily accessible at building material retailers, home planning centers, and at the building centers in most large department stores. You'll also find a wealth of planning information in consumer magazines and manufacturers' literature available where building materials are sold.

There are three basic steps in planning a plumbing project:

1. Select the proper fixtures. The first consideration is your need; the second is your budget. Plan with size, color, and styling of the fixtures in mind; you are going to live with them for some time. Also, think about the location of the fixtures in a room—bathroom, kitchen, laundry, basement. The fixtures should be convenient and easy to use. The fixtures also should fit your lifestyle. If they don't, you will be unhappy with them, and installing them will be a waste of time and money.
2. Estimate your plumbing needs. This involves determining the proper sizes and lengths of pipes, the types of fittings that are required for the pipes, and how and where the piping and drain connections will be made to your present plumbing system.
3. Install the plumbing. Or at least know enough about the plumbing plan in your home to help a professional plumber make the necessary changes. Check the plumbing codes in your area to make sure the piping and fixtures you plan to use meet the standards.

This chapter includes a potpourri of plumbing installation techniques. From the photographs and drawings you should be able to plan your plumbing project more efficiently and be able to do much of the work yourself.

Select the proper plumbing supplies

In planning a plumbing project, keep this information in mind:

There are six different types of pipes. Codes will dictate which type you must use.

1. Galvanized steel pipe. This pipe may be used in secondary drainage systems, but it is mainly used in water-supply systems along with regular fittings. Water-supply runs require ½-, ¾-, and 1-inch size pipe. You can buy the pipe already threaded, or you can thread it yourself.
2. Cast-iron pipe. This extremely durable pipe is the main component in waste disposal systems, particularly in underground installations. Fittings include gasket and no-hub connections.
3. Rigid copper pipe. This is available in ⅜-, ½-, ¾-, and 1-inch sizes. You can buy larger sizes for underground use. Only the ½- and ¾-inch sizes are usually used in home plumbing. Type L is recommended for interior plumbing runs. Use Type K for underground piping installations. Fittings are flared and

74

compression or sweat soldered.
4. Flexible copper tubing. This is available in 30- and 60-foot coils in two types, L and K. Flexible tubing's advantage is its ability to bend and turn corners. As with rigid copper pipes, the fittings are of the flared and compression type. Do not use flare-type fittings behind walls.
5. Rigid plastic pipe. Where codes permit, this material may be used for cold—never hot—water plumbing. Rigid plastic pipe may also be used for drains and vents, again depending on local codes. Rigid plastic pipe sizes correspond to galvanized steel and rigid copper pipe sizes. Fittings are solvent welded.
6. Flexible plastic pipe. This material is specified for underground outdoor use. It should carry cold water only.

Fixture water-supply lines are usually chrome plated. They are sold in rigid and flexible types, and the fittings needed to join the fixtures to the water supply are generally included.

There are three types of valves and faucets. The first is a gate valve used in water mains. Globe and ground-key valves are used inside the house. Stop and waste valves are globe or ground-key types; they are used for draining the shutoff side of a water-supply line.

Step-by-step planning

Any plumbing project should first be sketched on paper along with the proper dimensions. In this way you will have a better understanding of where pipe runs and connections are needed. You should do the work in sequence—first things first:
1. Hook up underground drains.
2. Rough in fixtures.
3. Put in main and secondary stacks.
4. Locate and install vent lines and branch drains.
5. Install piping for water-supply system.
6. Hook up fixtures.

Easy Home Plumbing is not designed to show the reader how to install an entire plumbing system. Rather, it is a guide to minor repair and remodeling jobs. The accompanying photographs should help familiarize you with plumbing designs and layouts you would not ordinarily be able to see.

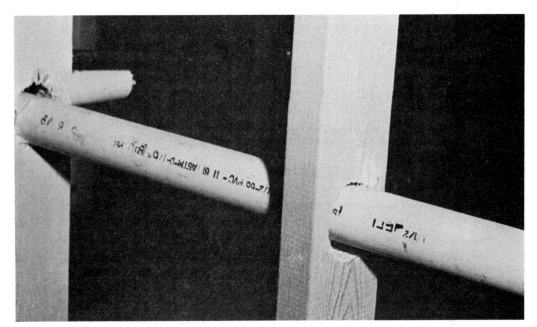

Pipe runs within walls go through holes drilled in the studs with an expansion bit or hole saw. Studs may also be notched to support the pipe. Fittings through which water flows must be the sanitary type. Horizontal water runs must have a 1-inch downward pitch every 4 to 5 feet to keep the water flowing.

Plumbing Projects 75

Studs at a corner are notched as shown to support the pipe run. Bore two holes, and connect the holes with a saw cut to form the notch. Notches in the sides of studs which carry straight pieces of pipe (not shown) should be reinforced with a piece of steel (metal mending plate) nailed or screwed to the studs in a vertical position to prevent the pipe from vibrating out of the notch.

Roughed-in plumbing for a kitchen sink looks like this. The water-supply lines (copper) are installed up through the floor; the waste system goes through studs.

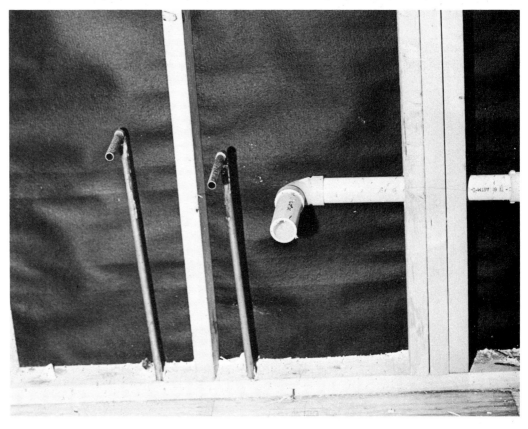

To go through concrete or concrete block with pipe you will need to drill a hole. Use a star drill or a masonry drill locked in the chuck of your portable electric drill. Since the pipes hug the wall, as shown, furring strips are needed to create space between the finished wall and foundation wall. Generally, 2 by 4s are used for furring. The 2 by 4s are set with the face of the material against the wall.

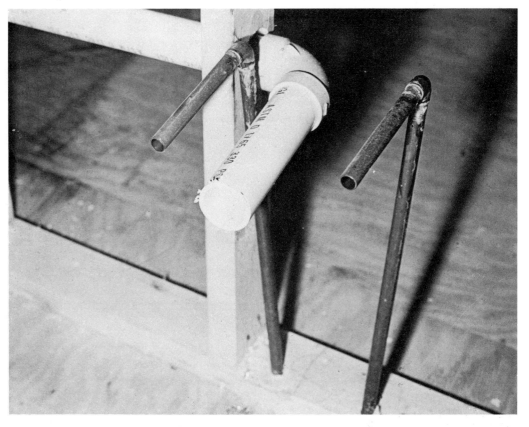

Lavatory plumbing rough-in. The drainage system is plastic; the water supply is copper tubing. The lavatory is held by hangers fastened to the wall. This support should be fastened to a 1- by 6-inch board that spans the studs.

Plumbing Projects 77

A bathtub sits directly on the subfloor. The flanges of the tub are supported by 2 by 4s at the front, back, and along the wall edge of the tub.

If the tub will be placed on a concrete floor, you generally have to leave space underneath for the drain pipes, as shown. Allow for this when fresh concrete is poured.

Bathtub supply lines extend into the wall as shown here. The studs are notched to recess the piping. When finished, the faucet assembly will be flush with the back of the wall.

Back of bathtub supply lines looks like this. The notch for the piping will be covered with a steel reinforcing (mending) plate. You should also have an access panel so that you can get to the fixture to service it. The access opening should be about 3 to 5 inches square.

Plumbing Projects

Shower head rough-in utilizes a 1- by 6-inch board to support the riser pipe. Set this board flush into the studs about 5 feet above the subfloor. Position the bathtub when the framing is ready and before the piping is installed.

Piping between joists. Shown is a plastic P-trap below the floor joists in a basement. If you are sweating copper joints around wood with a propane torch, be sure to shield the wood with a piece of asbestos board. Also, with copper and plastic pipe preassemble as much of the run as possible.

Toilet assembly rough-in includes a water-supply pipe for the flush tank and a plastic pipe for the waste. Install a shutoff valve on the water-supply line when you hook up the flush tank.

Laundry water supply. This new device teams the hot and cold water with a waste line. The connections may be made below the height of the washer and dryer so that the unit doesn't show and yet is very accessible for service.

Plumbing Projects 81

Notched furring strips permit enough room for a waste pipe and water-supply pipes. The furring here is 2 by 6s nailed to the floor joists and a plate fastened to the flooring above. Gypsum wallboard will go over the furring to complete the wall.

Vent stack (plastic) should extend about 12 inches above the top of the roof. It should be sealed with flashing that extends under the roof shingles. The joint should then be sealed with roofing cement.

Sink and dishwasher hookup

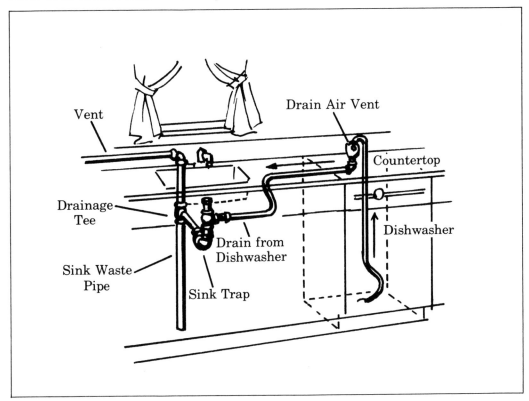

Sink with grease trap hookup

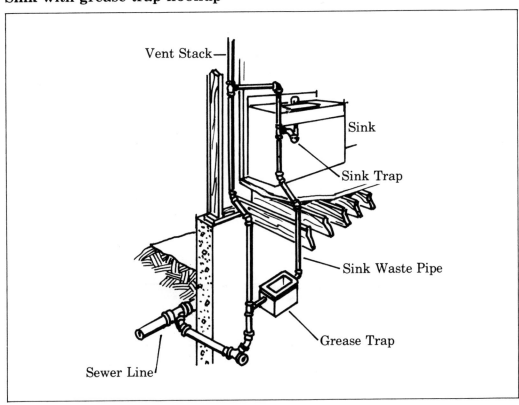

Water supply system for one bathroom and a kitchen

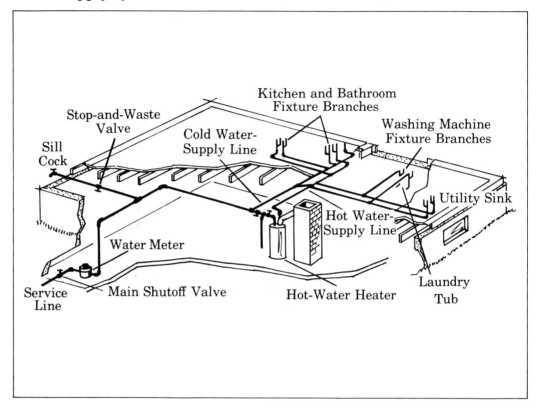

Two ways to join a new drainage system to an old one

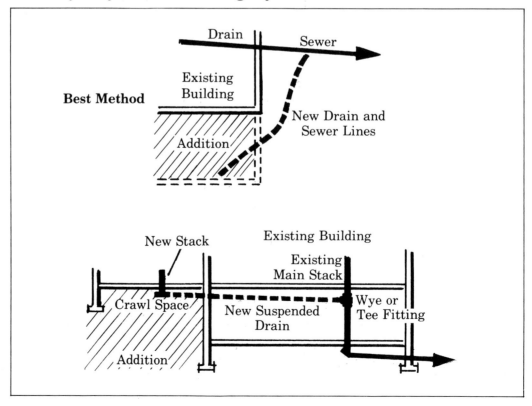

Reinforcing a cut joist with headers

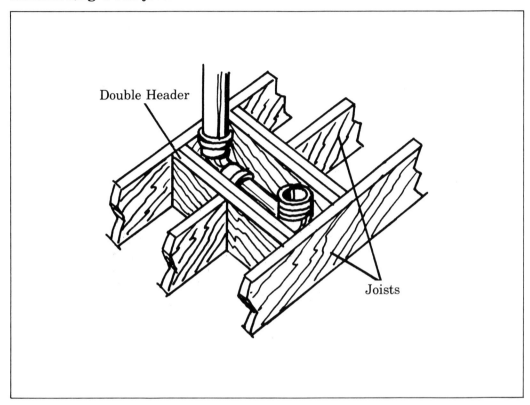

Horizontal pipes through joists

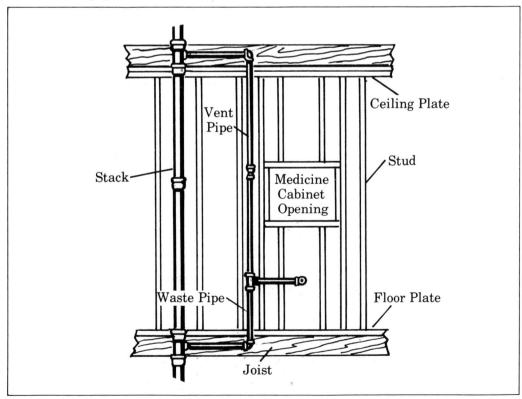

Running pipes through walls and floors

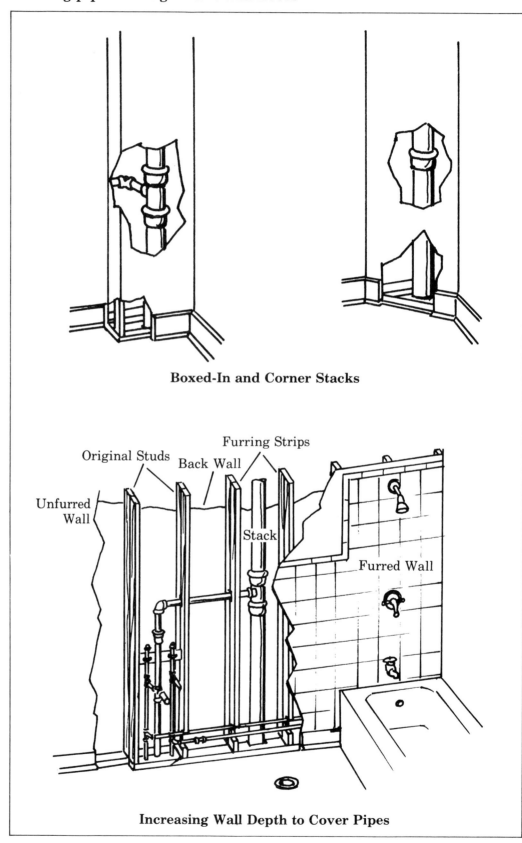

Boxed-In and Corner Stacks

Increasing Wall Depth to Cover Pipes

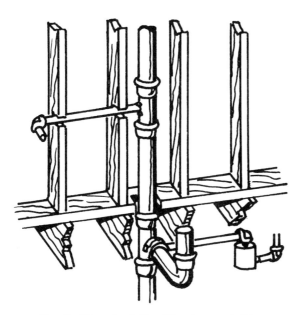

Studs Notched for Horizontal Pipe

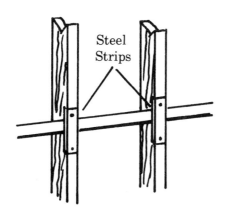

Reinforced Studs

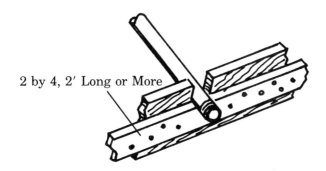

Reinforced Joists

Framing and measurements for a lavatory

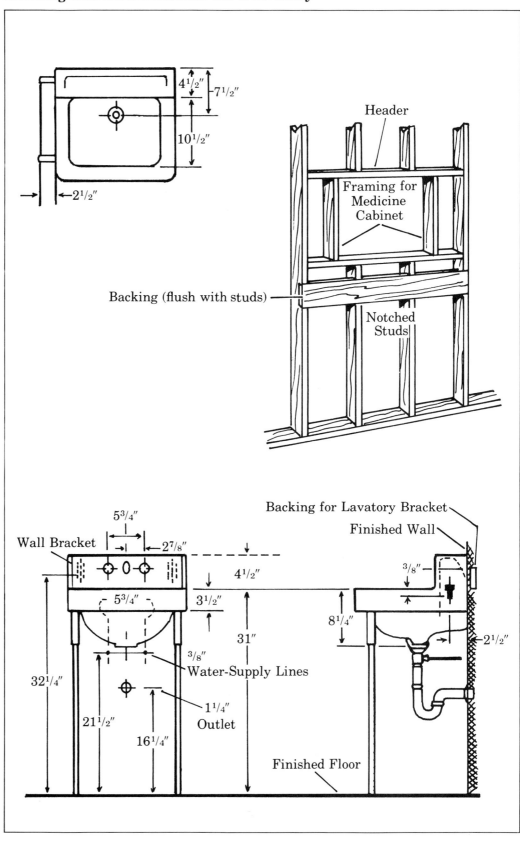

Framing for a bathtub

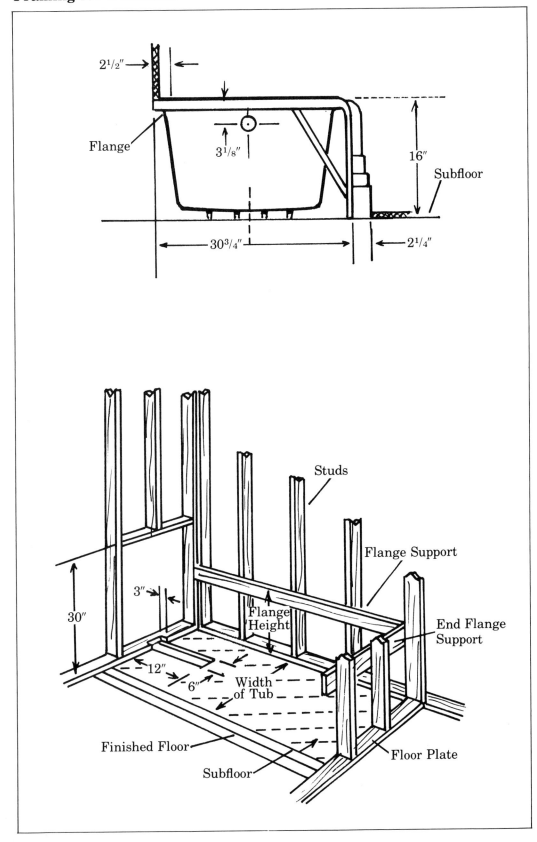

Framing for a floor-mounted toilet

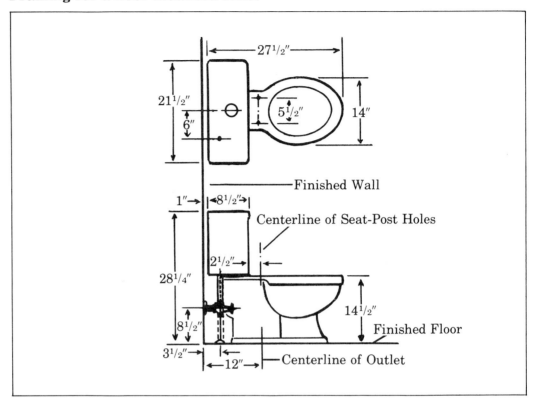

Vent stack detail

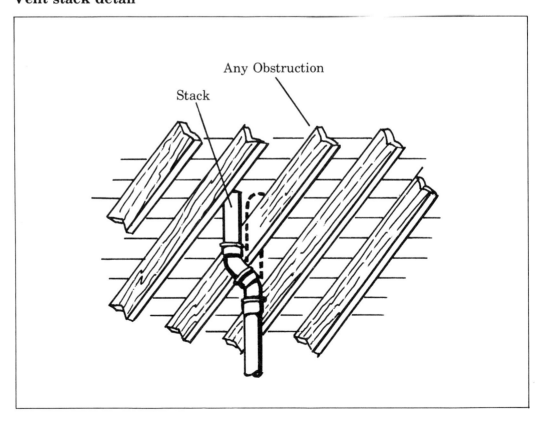

Locating an opening for a soil stack

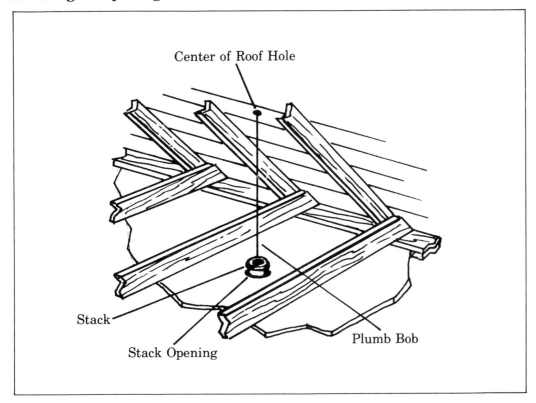

Typical waste and vent run

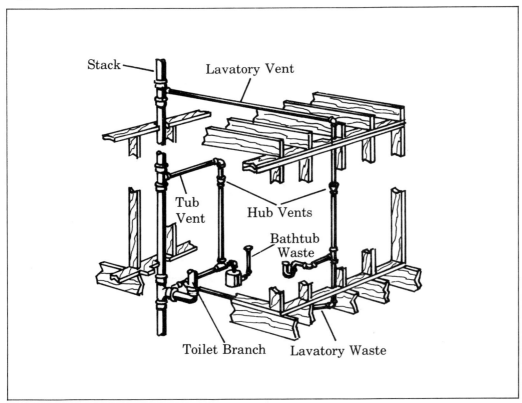

Connecting lavatory waste pipes

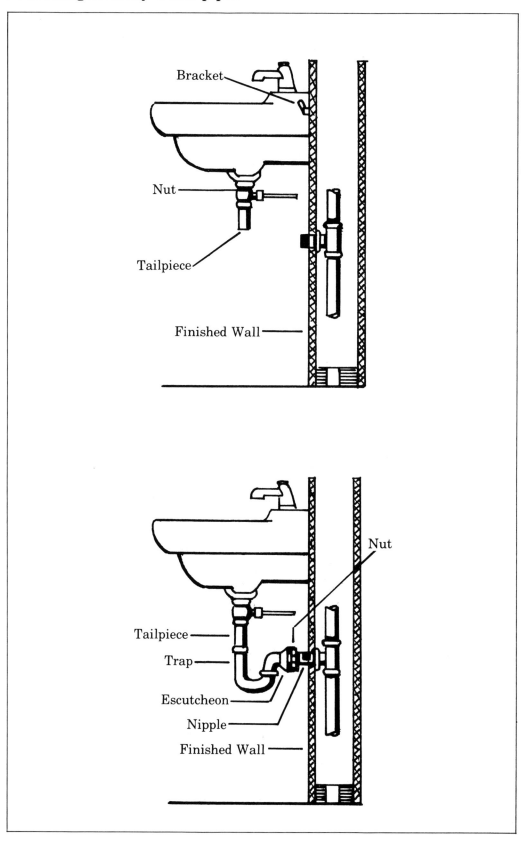

Two ways to connect a toilet tank

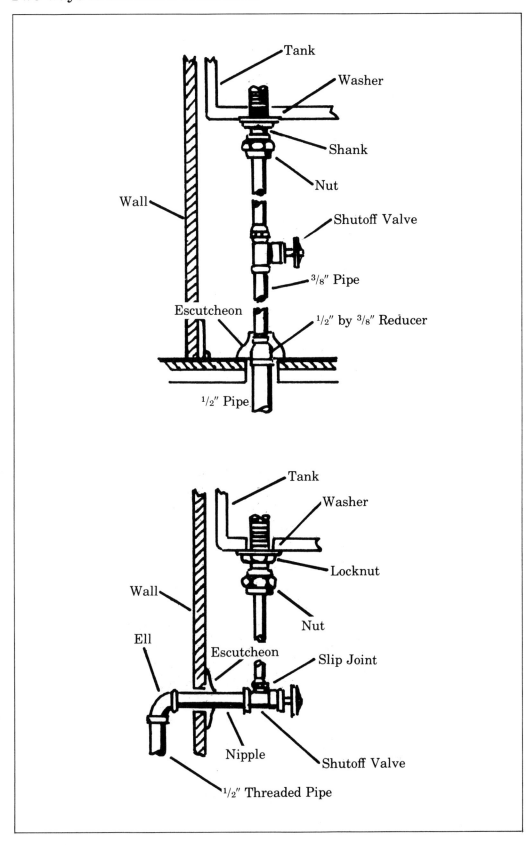

Measuring lengths of nipples

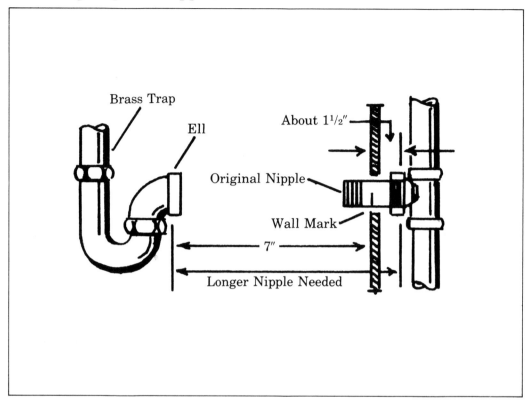

Setting a toilet bowl on its flange

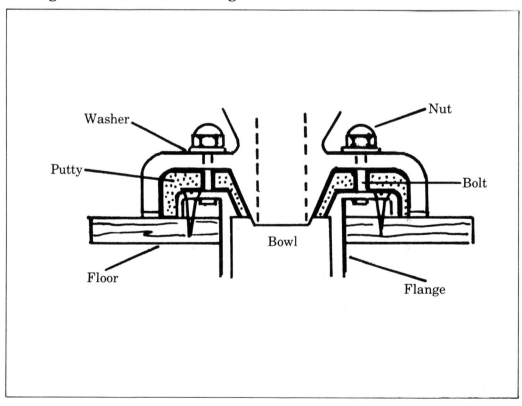

Index

A

A B C plastic pipe, 69
Adapter, 54
Air-cell asbestos, 30
Air chamber, 32
Angle bang, 32
Asbestos tape, 30
Assembling pipe
  cast-iron, 57
  steel, 50
Auger, 5, 10, 20–22, 33, 37, 38–39
Auto hose clamp, 27

B

Ballcock, fluid action, 45
Basic plumbing tools, 5
Bowl refill tube, 36
Branch drain, 36
Branches, 7
Brass fittings, 52
Bronze fittings, 52

C

Cast-iron pipe and fittings, 48, 57–58, 60, 69
Caulk, 58
Caulking, 55
C-clamps, 27
Cemented plastic pipe fittings, 72
Channel lock pliers, 9–11
Cleanout plug, 22
Clogged
  drains, 17
  toilets, 36
Compression fittings, 47, 61–62
  copper fittings, 52, 66
  copper pipe and tubing, 24, 61–68, 69
  nuts and rings, 5
Connecting lavatory wastepipes, 92
Copper tubing, bending, 66

CPVY plastic pipe, 64, 70
Cutting and assembling
  cast-iron pipe, 57
  copper tubing, 63
  plastic pipe and tubing, 70–73
Cutting oil, 50

D

Die, 67
Disc-type valve faucets, 15
Disposal
  drainage, 33–34
  field patterns, 60
Double wrench pipe assembly, 51
Drain fittings, 48, 52
Drainage
  disposal, 33
  hot-water heater, 33
  pipe, 57
  systems, 6–8, 48

E

Electrical heating tape, 28–29
Electrician's tape, 5, 24
External (male) threads, 50–51, 56

F

Faucet
  aerator, 23
  chatter, 32
  leaking, 9
  single-handed, 14
  slow-running, 16
Female threads. *See* Internal threads
Fiber glass insulation, 30
Fiber pipe, 48
Fixing leaky pipes, 24
Fixtures, 7
Flared
  and compression copper fittings, 66
  copper joint, 67–68
  copper tubing, 5
  fittings, copper pipe, 61–62

Flaring tool, 61, 67–68
Flexible
  copper tubing, 60–62
  plastic pipe, 73
  plastic tubing, 69, 72–73
    fastened to fittings, 73
Float arm, 39–40
Float ball, 36, 40–42
Flush
  handle, 43
  tank, 36, 40–42, 45, 47
  cracked, 36
  mechanism, 38
  problems, 39
Flushing lever, 36
Framing
  bathtub, 89
  floor-mounted toilet, 90
Frozen pipes, 25, 28

G

Galvanized pipe, 5
  steel, 24
  steel pipe and fittings, 48–60
Garbage disposals, 9–10, 33
Graphite packing, 5, 9, 13
Guide arm, 42

H

Hanger brackets, 32–33
Horizontal pipes through joists, 85
Hot-water heater, 5, 33–35
  drainage, 33
  thermostat, 33
Hub joint (cast-iron), 58
  and spigot cast-iron pipe and fitting 5, 55

I

Insulating material, 25
Insulation, 30
Internal (female) threads, 51, 56

J

Joint runner, 57

L

Leaking
  faucet, 9
  hot-water heater, 34
  pipes, 24
  toilet tank, 45
Lifting wires, 39, 42, 44
Lime, 9, 24

M

Main sewer line, 22
Main shutoff valve, 5, 24, 27
  water valve, 6
Male threads. *See* External threads
Mica-type insulation, 25
Molten lead, 58

N

Neoprene gaskets, 57
No-drip tape, 30–31
No-hub cast-iron pipe and fittings, 56
  joints, stainless steel shield, 59
Noises in pipes
  angle bang, 32
  faucet chatter, 32
  water hammer, 32

O

Oakum, 58
O-rings, 9, 44
  washers, 9, 44
Overflow outlet, 19, 24
  tube, 39, 41

P

Patches, 25
Pipe clamp, 26
  compound, 24
  cutters, 49
  jackets, 30
  joint compound, 51, 53
  length, 49
  reamer, 50

Plastic fittings, 52, 70
  foam, 30
  pipe, 24–25, 48
    and tubing, 69–73
  tape, 51
Plumber's caulking iron, 58
  putty, 34
Plumbing codes, 48, 61
  maintenance, 9
  terminology, 7
  tools, 5
Plunger, 5, 19–22, 36
Propane torch, 29, 61, 65
PSI plastic pipe, 69
PVC plastic pipe, 69

R

Regular threaded fittings, 48
Rigid copper pipe, 60
  plastic pipe, 69, 70
Risers, 7, 32
Rubber flapper, 36
  gasket, 26–27, 46

S

Sanitary drainage fittings, 48
Self-tapping screw, 28
Septic system, 59
Shutoff valves, 7, 25
Sill cocks, 25
Sink flange, 34
Soil, stack, 6, 36, 48
Spigot, 57
Squeaking pipes, 32
Standard fittings, 48–49, 52
Steel fittings, 56, 73
Stem assembly, 9–15
Stoppers, 17–19
Strainer, 15–16
Sweat soldered copper fittings, 61–66

T

T-fitting, 32
Threadcutter assembly, 50
Threaded fittings, 48

Toilet, cracks in, 46
  difficulty in flushing, 43
  inlet valve assembly, 44
  valve seat, 45
Tongue wrench, 59
Traps, 7, 20–22, 36
Tube cutter, 5, 61, 63

U

Union fittings, 52–54

V

Valve grinding tool, 13
  seat, 13, 16
Valves, 7
Vents, 6–7
  fittings, 48, 52

W

Washers, 9, 12, 45
Waste pipes, 5, 7
Water hammer, 32
Water pipes, 5
Water-supply inlet valve, 36
  line, 32
  pipes, 5–6, 8
  systems, 48